Reviews

This outstanding book presents an impressive collection of surprising facts about the brain, the miracles it creates, but also the mistakes it makes. The author explains that the positive achievements and negative outcomes are produced by conscious and subconscious models that the brain constructs and operates. This formulation is consistent with the most recent results of brain science. From this concept, he is even able to derive philosophical conclusions such as "the concept of free will is nonsensical." Fascinating!

—**Martin Grötschel, mathematician and former President, Berlin-Brandenburg Academy of Sciences and Humanities, Germany**

A novel way to look at the human mind. Magic, Error, and Terror is a good title for this insightful analysis of the mind based on his models of the conscious and unconscious. I learned from Wittgenstein that if you cannot give examples in an abstract discussion, you are talking nonsense. I like this book because the author gives a lot of examples from real life situations. For example his discussion of fatigue and panic attacks and their treatment lets you have a grip of the magic of his models. His analysis of medications given to postmenopausal women and religion as examples of errors and terrors of models are enlightening. The book is written in a clear and accessible fashion. I highly recommend it.

—**K. T. Fann, former professor of philosophy, York University, Canada. Author and editor of several books on philosophy**

A veritable scholarly "page turner!" Starting with the very first chapter of the book the author captures your attention, building anticipation for the chapters to follow. This book is concise, clearly written and opens our understanding to new ways of interpreting reality. It is both hopeful and terrifying! A truly engrossing piece of work!

—**Michele K. Steigleder, Ph.D., Clinical Psychologist**

Also by Klaus Truemper

Brain Science

Artificial Intelligence
Wittgenstein and Brain Science

History

The Daring Invention of Logarithm Tables
The Construction of Mathematics

Technical

Logic-based Intelligent Systems
Effective Logic Computation
Matroid Theory

Edited by Ingrid and Klaus Truemper

F. Hülster *Introduction to Wittgenstein's*
Tractatus Logico-Philosophicus
(English and German edition)

F. Hülster *Berlin 1945: Surviving the Collapse*

MAGIC, ERROR, AND TERROR

HOW MODELS IN OUR BRAIN SUCCEED AND FAIL

KLAUS TRUEMPER

Leibniz Company

Softcover published by Leibniz Company
2304 Cliffside Drive
Plano, Texas, 75023
USA

Original edition 2021
Updated edition 2021

The book is typeset in LaTeX using the Tufte-style book class, which was inspired by the work of Edward R. Tufte and Richard Feynman.

Sources and licenses for all figures are listed in the Notes section.

Library of Congress Cataloging-in-Publication Data
Truemper, Klaus, 1942–

Magic, Error, and Terror: How Models in Our Brain Succeed and Fail
Includes bibliographical references and subject index.
ISBN 978-0-9991402-2-2
1. Brain. 2. Model.

Contents

1

Introduction

The human brain[1] has enabled mankind to

- create bountiful harvests
- design powerful computing machines
- conquer diseases once thought incurable
- replace body parts
- fly to the moon and back

We could go on and produce a long list of near-miracles. What made this possible? Two factors come to mind.

- The brain develops its basic capabilities in infancy and later changes itself as the need arises.[2] The end effect is an incredible performance of complicated tasks—like the processing of images and language—that our fastest computers can't match.

- The brain builds and stores in its various parts complex *models* that represent features and processes not just of the world, but also of the body.[3] These models are essential for interaction with the world and management of the body.[4]

All this by an organ weighing less than 4 lbs.[5]

————————————

Stephen Hawking's book *The Grand Design*[6] formalizes the concept of models inside the brain.

It declares that some of the models are stashed away in the brain below the level of consciousness. They are *subconscious models.* Their defining characteristic is that we aren't aware of them in our daily lives.

Other models reside at the level of consciousness. They are sometimes written down and stored in books and technical journals, or more recently on the Internet. They are *conscious models* since they are easily accessible and we can readily describe them. They include conceptual models such as Newton's model of gravity and the Big Bang theory of the universe's origin.

———————

The idea of subconscious and conscious models is related to, but not the same as, the notion of intuitive and deliberative thinking.[7]

Yes, subconscious models sometimes produce results that appear at the level of consciousness as intuition. In fact, one could characterize intuitive thinking that way. Also, conscious models are part of deliberative thinking.

But the concept of subconscious models encompasses much more. Let's rank them according to the degree of difficulty with which they can be accessed by conscious thought.

At one end of the scale are subconscious models that can be accessed with a bit of effort. For example, we may become fearful when we see dogs. Thinking about that reaction and searching our memories, we recall that a dog bit us decades ago.

Next are models we cannot identify readily, but are able to discern by significant effort. An example are subconscious models discovered during psychotherapy.

Finally, there are subconscious models that are essentially inaccessible.[8] An example is the subconscious processing of visual information, such as identification of the boundaries of objects. In a more complicated case, a subconscious model determines that on-going, strenuous physical activity would damage the body if continued not just for hours but for days. The model output triggers a feeling at the conscious level that rest is required and one needs to sit down.[9]

You may be surprised that we use the single label "subconscious" for such a variety of models. The alternative would be a classification using several terms. This would be a formidable task, given the complexity of the possible cases. It also would invite futile discussions whether specific cases have been appropriately classified.

Our single label avoids such debates and allows us to focus on the role, interaction, and impact of these models in various settings.

For the same reason, we generally won't attempt to specify where in the brain the various subconscious and conscious models reside and how communication between models is accomplished.

———————

In the night sky, some bright points of light seem to always move together, while others wander individually. Extensive research carried out over centuries resulted in a conscious model that not only classifies the two types of lights as stars and planets, but predicts with high precision their movement.[10] In particular, the model predicts that the planets move around the sun in elliptical orbits.

The results of the model are often cited as if they were facts of the world: The orbits of the planets around the sun are then claimed *to be* elliptical.

But that is not the case: The planets influence each other in ways that have proved to be mathematically intractable: While approximate formulas predict the movement for limited time intervals, we can only guess what will happen long-term.[11]

The example isn't an unusual case of model interpretation: We conflate model output with facts of the world on a huge scale.

Indeed, much of what we decide and do is based on this assumption. Appropriately, in the book *The Grand Design* this conflation is called *model-dependent realism*.[12]

Here are the key components.

- The brain uses a large variety of subconscious and conscious models to cope with the world.
- The brain declares the output of these models to be facts.
- The brain bases decisions on these supposed facts and considers them valid since they are, well, fact-based.

Fortunately for us, the declaration of model output as fact mostly has good if not wonderful consequences. At times the results look like magic. Examples are today's computers, communication devices, the Internet, and the Global Positioning System (GPS).

The key word in the preceding paragraph is "mostly." There are exceptions where model-dependent realism prevents important developments, causes harm, or results in utter destruction and terror.

We can prevent such negative or even horrific results when we acknowledge that we are dealing with model output and not with facts, and then modify or replace undesirable models by others. That way the conflation of model output with facts no longer causes harm.

The prevention invariably entails that we first acknowledge that we are dealing with model output and not facts. After that step we are ready to improve or replace models.

We may adjust or replace not only conscious models, but also subconscious ones. Many psychotherapy interventions can be viewed as replacement processes of the latter type.

In the chapters that follow, we look at diverse scenarios where model-dependent realism produces desirable results, or harm, or even terror. We also see how a shift to different models eliminates some cases of the latter two kinds.

We say "some cases" since an outcome may be the sum of effects produced by a model residing in millions if not billions of brains of the people of an entire country or even the world. In that case, us changing the model in our brain has no noticeable impact.

Examples are the human overpopulation of the earth created by the drive of most couples to have babies, and the destruction of the environment and mass extinction of animals due to the human urge for ever-increasing consumption of resources.

A great many books propose new models that, if installed in most brains of the world, would change these overall outcomes. But the books fall woefully short on realistic recommendations how people can be convinced to adopt these models. Evidence of failing installation attempts are the fruitless debates about the wholesale destruction of the earth's lands and oceans, and the mass extinction of animals. If we were to ask the dolphins, whales, and elephants about this, they would say that the earth suffers from a pandemic infestation by humans.

We will not add to these proposals here since we are rather pessimistic about the installation attempts. We just hope that humanity will eventually find a way to avoid looming self-destruction.[13]

———————————

The book consists of three distinct parts that cover a kaleidoscope of cases. They demonstrate the universal role of subconscious and conscious models in our lives.

- Part I tells how some subconscious models can be changed. It is by no means self-evident that this can be done, since by definition subconscious models are not easily accessible by conscious thought if at all. The examples cover psychotherapy, the notion of fatigue, the control of breathing, optimal motion, relaxation, and training in general.

- Part II covers deceptive conscious models of medicine, economics, politics, and religion. We focus on deceptive conscious models since there is a huge literature on successful ones.

- Part III uses conscious and subconscious models to investigate statements of philosophy. In particular, we look at the question whether we have free will.

Even if you are not so fond of philosophical discussions, you may still want to look at the first two chapters of Part III since they describe a method for identifying nonsensical questions and statements. We use it almost daily to separate useful grain of news reports from chaff.

As you read the chapters, you surely will come up with more ideas about subconscious and conscious models; their impact; and how to create, modify, or replace them. For example, you might consider addiction to be an instance of deceptive subconscious models, and view treatment as modification of such models. Part I contains relevant material for such an investigation.

Stimulating such thoughts is another purpose of the book.

Let's get started with Part I: How can we modify subconscious models?

Part I

Changing Subconscious Models

2

Things We Will Never Learn

In March of the year 43 BCE, Marcus Tullius Cicero (106–43 BCE) gave a long speech in the Roman Senate that included the statement,[14] "Any man is liable to [make] a mistake; but no one but a downright fool will persist in error."

As with so many sayings of famous thinkers, it seems utterly true: Why would any sane person repeat a mistake?

But the statement is quite false: There are instances where a sane person repeats an error over and over, all the while knowing that the action is a mistake. Sounds crazy, doesn't it? But it is true, nevertheless.

Let's look at an example where we know that a decision is wrong, yet cannot avoid that choice.

We are shown the picture below and are asked about the color of the two squares labeled "A" in the topmost row and "B" next to the cylinder. More specifically, how do the pixels of squares A and B differ?

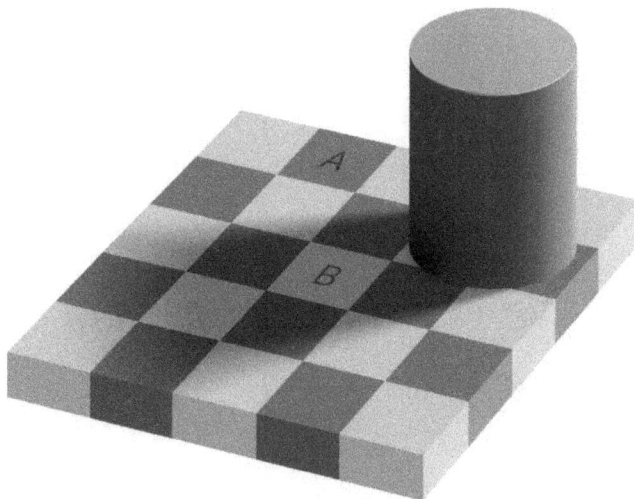

Perception: Square A is much darker than square B.[15]

The answer is obvious: The pixels of A are black and those of B are white.

If we want to be more precise and avoid the extreme terms "black" and "white," we may instead declare that the pixels of A are much darker than those of B.

The brain's computation of that decision is complicated. But summarily we may say that certain processing starts at the subconscious level with the visual impression recorded by the retina of the eyes and eventually terminates with the final decision at the conscious level.

We could say that some subconscious models are part of the evaluation process, such as identification of the border of the checkered piece and of the column.

Next we are shown a quite similar picture.

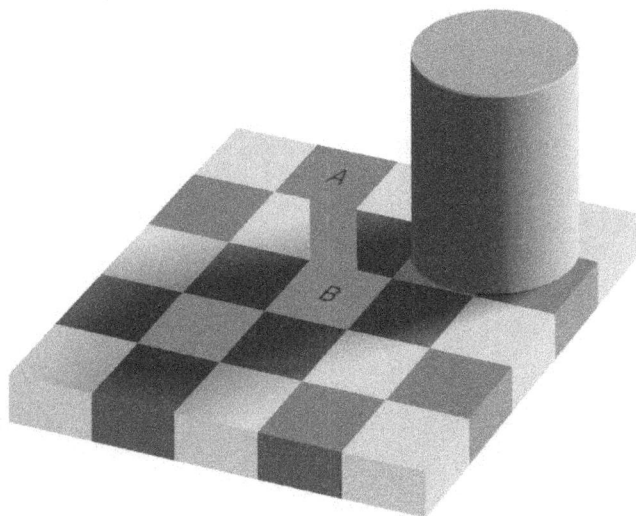

Reality: Squares A and B have the same shade of gray.[16]

The only change is that the squares A and B have been connected by a corridor that has exactly the same gray shade as both A and B. We suddenly realize that our answer has been wrong.

Incredulous, we go back to the first picture. Once more we become certain that A and B are differently colored. Moving then to the second picture, we see again that this conclusion is wrong. In fact, we can go back and forth as often as we like and each time experience this effect.

What is happening here?

In the first picture, the brain uses a subconscious model to determine color where the shading effect of the column is taken into account.

Since this occurs at a level below consciousness, we have no control over it. Indeed, no matter how we try to counteract this interpretation after seeing the second photo, the brain will insist on that interpretation for the first picture.

For the evaluation of the second picture the brain uses a subconscious model that essentially says that a contiguous area that looks uniformly colored is indeed uniformly colored. That model is used to evaluate the area composed of A, B, and the connecting corridor.

The amazing aspect is that our brain consciously knows that a mistake is being made, but cannot correct the subconscious model to avoid the problem. Indeed, the incorrect model in some sense is hard-wired and cannot be modified.

Here is another case where a decision is wrong, yet we make that same choice once we have become aware of the error.

How do the two parallelograms in the drawing below differ?

Two parallelograms as table surfaces.[17]

The answer is obvious: The left parallelogram is much longer and slimmer than the right one.

Now strip away the legs and moldings of the tables.

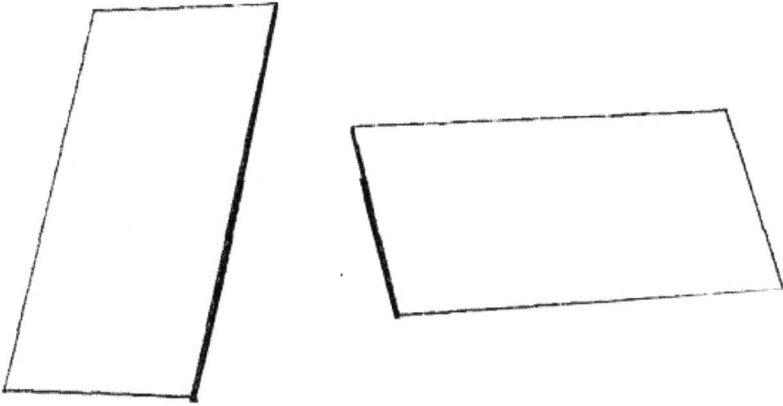

The two parallelograms seen by themselves.[18]

The right parallelogram is now seen as a rotated version of the left one.[19] We confirm this by measuring with a ruler.

Given that knowledge, we go back to the first picture: The two parallelograms become different again. Indeed, our brain cannot resolve this conflict, just as it couldn't for the two squares A and B of the checkered surface. Apparently the subconscious model cannot be changed.

———

Is it then true that we cannot ever change *any* subconscious model by conscious actions and thoughts? Fortunately, the answer is an emphatic "No." The subsequent chapters demonstrate this. We begin with a case where change is quite difficult.

3
Psychotherapy

Psychotherapy relies on a variety of models to treat mental illness. We look here at one particular approach: *cognitive behavioral therapy (CBT)*.

Extensive research has demonstrated CBT to be an effective treatment for a range of emotional problems such as anxiety, depression, and panic attacks.

The key idea of CBT is:[20] *Our thoughts trigger our feelings; therefore changing our thoughts will change our feelings.*

The notion seems simplistic, doesn't it? It is anything but.

We cover the key ideas of CBT with the story of a fictitious female patient named Sophia. Her thoughts and actions are based on actual cases.[21]

We will meet two models: The first one provides an understanding of the situation at hand, while the second one is the basis for treatment.

———————

Sophia is reaching 50 years of age.

This is about the time when her body begins to slow down a bit, and it takes more effort to work at her accustomed pace.

Simultaneously, the satisfaction derived from work is eroding. So at the end of each day, the thought "I didn't achieve much today, did I?" pops up. At the same time she feels tired.

So there are now two conflicting feelings: First, that she has not done enough work, and second, that she is tired and should stop.

Sophia consciously decides to work more and ignore the fatigue. In some sense, she is sacrificing the well-being of the body to achieve the satisfaction of accomplishing more work.

A CBT therapist would classify the thought "I didn't achieve much today, did I?"as being generated out of awareness, and the fatigue to be a reaction of the body to the work effort.

Here, we interpret the process using the concept of subconscious models defined in Chapter 1.

We say that Sophia has a subconscious model evaluating her work performance. She may have adopted that model decades ago while listening to her parents or teachers. Regardless of the case, the model judges her performance and triggers the thought "I didn't achieve much today, did I?"

As we will see in Chapter 4, another subconscious model produces the feeling of fatigue.

So now there are two conflicting feelings that the conscious brain needs to deal with. Sophia resolves them by working more and ignoring the fatigue.

The process repeats day after day. Each time the conscious brain decides that the satisfaction of more work is more important than the feeling of fatigue.

Eventually the part of the brain operating the subconscious fatigue model is fed up by Sophia's behavior. In an emergency action it throws a wrench into the machinery. This can take various forms.

In Sophia's case, the intervention is a simulated heart attack. That is, the symptoms are there: chest pain, a feeling of weakness, shortness of breath, rapid pulse. But actually, there is no heart attack.

The conscious brain is alarmed and reacts by calling for an ambulance. The emergency medical technicians see the symptoms, decide that there has been a heart attack, and bring Sophia to the hospital.

The doctor first believes there has indeed been a heart attack, but after some tests concludes that this wasn't the case.

Medical doctors are not trained to diagnose the cause of this false heart attack, so Sophia stays overnight for observation and is sent home the next morning. All is well, or so it seems. After all, the doctor said so.

Sophia now makes a big mistake. Since there was no heart attack, she goes back to her accustomed work habits, where the conscious brain decides to do more work and ignores fatigue.

Four weeks later, there is another supposed heart attack, followed by a trip to the hospital and the eventual diagnosis that this heart attack is a false alarm, too.

Sophia sees the family physician, who recommends psychotherapy. Sophia visits a psychotherapist, who explains that she is having panic attacks. They meet weekly for treatment.

By this time, one might say that the subconscious brain is totally fed up with Sophia's behavior. Indeed, the conscious brain has become a real threat for the well-being of the body. Thus, the subconscious part starts panic attacks when there is the slightest perception that the body is being overtaxed.

The panic attacks are no longer confined to false heart attacks. Here are some examples.

- Sophia feels that her heart has stopped beating. She is convinced that she will die within a few minutes since the brain is no longer supplied with oxygen.
- She becomes so dizzy that she has to sit down immediately and does not dare to get up again.
- She has heart palpitations: Her heart beats fast, flutters, or pounds.
- She thinks she is going crazy.
- She is utterly convinced that her blood pressure has gone down to zero.
- She suddenly thinks that death is imminent.

Each frightening instance is accompanied by feelings of utter helplessness and loss of control.

Altogether, the panic attacks achieve their common goal: Sophia slows down and reduces her demands on the body. Thus, from the viewpoint of the subconscious brain, the method is working.

––––––––––––––

The therapist must accomplish the following:

- The conscious part of Sophia's brain must be persuaded that a much more gentle treatment of the body is needed.

  This is not so difficult: The therapist suggests leisurely walks, where the focus is to observe nature; meditation, such as mindfulness; relaxing activities, like painting or playing music; and, of course, a reduction of the workload.

  Since the proposed changes are decided by the conscious part of Sophia's brain, she only needs to listen to the therapist, accept the proposals, and implement them.
- The subconscious brain must be convinced that there will be no more excessive work damaging the body.

  Do you see the difficulty? The subconscious brain uses a model that determines when the body is being overtaxed. That model is triggering the panic attacks. It must be replaced by one that concludes that the reduced work load is okay for the body and that the new relaxation activities are helping to restore it.

The second item sounds like an impossible task, doesn't it?

But it can be done: Sophia learns conscious thinking that effects gradual replacement of the old model in the subconscious brain by a new one.

Specifically, Sophia learns to debunk the main claim of each panic attack by:

- *Understanding the faulty logic of the panic attack*: "The heart cannot possibly have stopped."
- *Decatastrophizing the main claim of the panic attack* : "So what if I become dizzy? People will help me."
- *Reattributing the supposed reason for the panic attack*: "I just feel hot and sweaty because of exercise, not because the heart is failing."

These coping mechanisms are supplemented by the following, general thoughts.

- Preparing as a panic attack begins:
  "I know that heart and body are in excellent condition."
  "I must continue with whatever I am doing, so that I will learn how to cope with fear."
  "If I give in, it will only get worse."
- Confronting and handling attacks:
  "I will proceed deliberately with my current task, knowing that heart and body are alright."
  "There is nothing wrong physically."
  "It will become easier as I do this time and again."
  "It is just an erroneous interpretation of signals that is causing the anxiety."
- Coping with feelings of being overwhelmed:
  "I know that the fear may increase. But physically I am still alright."

"If I go on, the fear eventually will subside since heart and body are not overloaded."

"I can do this, so I will just go on."

"The fear will not harm heart or body."

"No matter what, the fear will eventually subside. I just need to stick to the plan."

- Pleasant thoughts after successful handling of an attack:

"It worked: I did it!"

"When I controlled my negative thoughts, I also controlled my anxiety."

"This successful outcome will help me later."

"I really like how I am making progress."

"I am really learning how to conquer my fears."

Altogether, these thoughts and coping mechanisms are like tools in a portable toolbox. No matter where Sophia goes, she has these tools available and ready for use. This gives Sophia the confidence that she will be able to cope with any and all attacks, no matter when and where they occur.

As she employs these tools, the frequency and length of the panic attacks gradually lessen. In terms of models, the old reaction of the subconscious part of the brain is gradually replaced by a more realistic one where normal activities are no longer considered an assault on the body.

As the panic attacks subside, so do the opportunities for learning how to cope with them. Sophia then looks for panic attack triggers and initiates them voluntarily for additional *exposure treatment*.

Here is an example. She has observed that, when she exhales deeply and waits just a few seconds before inhaling, the panic attack "My heart has stopped beating" comes on.

So now she exhales, waits until the panic attack starts, holds as long as she can bear the thought that her heart has stopped, and then inhales. Whereupon the panic attack disappears instantly.

While the panic attack is building, she goes through the thoughts listed above: "I know that my heart is okay. There is nothing wrong physically. It is just an erroneous interpretation of signals that is causing the anxiety."

Doing this exercise every day, the onset of the panic attack is delayed further and further. After a while, it becomes impossible to trigger the panic attack that way.

The length and severity of the attacks are reduced in such small decrements that, going from one day to the next, there seems to be no change.

To make the progress visible, she keeps a log of all attacks and their severity. Looking over weeks of the log, she sees that things are getting better. This triggers positive thoughts that reinforce the change process.

And thus, over time, the panic attacks occur less and less often and are shorter and shorter, and eventually become just a nuisance.

Once that point is reached, she thinks at the onset of an attack, "I know you. You are just a panic attack. There is no need to worry. We will just go on with our work, and you will disappear." And then the panic attack subsides.

Eventually, the panic attacks disappear entirely.

At that point she may be tempted to use the tools of CBT to outfox the subconscious brain, so to speak. She then works just as hard as before and combats any panic attacks with CBT. For good measure, she throws in meditation whenever exhaustion sets in, then continues work as soon as the body has relaxed.

Eventually, these actions will cause the body to cave in, and what before was a *false* heart attack will become the *real* thing.

Thus, Sophia must not only overcome the panic attacks, but must also change her lifestyle to reduce stressful work and replace it with relaxing activities. This sounds like an easy task, but actually is difficult since pleasures derived from stressful work—"Isn't it great that I could solve that difficult problem so fast"—are being eliminated.[22] The changes may even result in depression. Here, too, CBT can help to right the ship of life.[23]

We have seen how CBT is based on an astonishingly simple, yet very effective idea where thoughts change feelings.

In the terminology of models used here, CBT uses thoughts at the conscious level to replace subconscious models.

Other interventions of psychotherapy are based on different ideas. Common to them is the use of models that explain human behavior and are the basis for treatment.

So far we have not mentioned the use of drugs as part of the treatment. They play an important role. Generally speaking, they are most effective when they are part of psychotherapy.

Lastly, some of the models of psychotherapy are effective when the patient just reads about them and has limited contact with a psychotherapist via videoconferencing, or just over the phone. Such treatment is called *bibliotherapy*. CBT has proved to be effective in that setting, for example for the treatment of mild to moderate depression.[24]

The next chapter covers the model for fatigue mentioned earlier.

4

Fatigue

We are hiking in the mountains: We climb up slopes and descend into valleys, all the time enjoying the scenery. After four hours, we feel tired and decide to rest.

Where does this feeling come from?

The obvious explanation is: The leg muscles determine that they have been stressed and are tired. They send a signal to the subconscious brain that they need rest. The subconscious brain then outputs to the conscious brain the feeling "The body is getting tired, and we need to take a break."

Suppose that explanation is correct, and a statement such as "My legs are getting tired" is an appropriate description of the situation.

How then is the following possible?

In 1986, Georges Holtyzer of Belgium walked 418 miles in six and a half days. He was not permitted any stops for rest and moved almost 99 percent of the time.[25]

Why do we get a feeling of fatigue after four hours of hiking when Holtyzer could walk more than six days without rest?

We get an answer by addressing the following, more fundamental question:

Is fatigue produced by our mind or is it signaled by the body?

Research into the causes of fatigue started in the 19th century. It led to the explanation that lack of oxygen and build-up of lactate caused muscle fatigue. Exercise textbooks from the 1930s to today advance this theory.[26]

Here are some problems with that notion:[27]

- Even at peak exertion, only about two-thirds of available muscle fibers are active.
- The feeling of fatigue is delayed when music is played during the activity—as is invariably done at exercise clubs around the world.
- When a wall clock is slowed down, people become tired later.

So something else must be happening. The crucial insight came during the 2010s, when fatigue was recognized as an emotion:[28]

The feeling of fatigue originates in the subconscious brain to ensure that ongoing physical efforts don't overtax the body.

So here is the correct explanation for the production of fatigue:

A model in the subconscious brain estimates whether the current effort if continued not just for hours but days would damage the body.

Once the brain comes to that conclusion, it sends a feeling of fatigue to the conscious brain. That feeling prods the person to rest.

Thus, fatigue is an emotion that protects the body from harm.

Evidently, the signal of fatigue is based on an extremely cautious evaluation. Athletes competing in long-distance races eliminate that unjustified evaluation with proactive thoughts that modify the fatigue model in the subconscious brain.

Such replacement is not easy since the model is not directly accessible: The runner can only engage in thoughts that gradually replace the restrictive model by a different model counseling perseverance.

The famous Finnish distance runner Paavo Nurmi put it thus:[29]

"Mind is everything. Muscles are pieces of rubber. All that I am, I am because of my mind."

Doesn't that explanation remind you of the approach of cognitive behavioral therapy (CBT) of Chapter 3? There, thoughts overcome panic attacks, while here they conquer feelings of fatigue.

There is another parallel. Remember that Sophia of Chapter 3 may be tempted to use CBT to hang onto her old lifestyle, thus setting herself up for a real heart attack?

The analogous situation applies to runners. Their thoughts can re-place feelings of fatigue by feelings of perseverance until the body reaches its physical limit. This happens in one of two ways:[30]

- The runner has calibrated the effort so carefully that she is at the point of physical exhaustion just as she reaches the finish line. At that moment, perseverance is overtaken by fatigue, and the runner collapses. But minutes later, she is up again and celebrating.

- If the runner has miscalculated and exhaustion sets in before the end of the race, that calamity manifests itself in the "Full Foster" collapse position[31] where the runner crawls on elbows and knees and finally collapses before or after reaching the finish line. Regardless of the case, survival is threatened.

The next chapter has another case of flawed interaction of subcon-scious and conscious thinking with the body. Once more, the main idea of CBT applies: Conscious thoughts insert the correct model into the subconscious brain.

5

Breathless

Chronic obstructive pulmonary disease (COPD) is characterized by breathing problems and poor airflow.[32] The most common symptoms are shortness of breath and a cough that produces sputum.

The disease progresses over time. At present there is no known cure, but the symptoms can be treated and the progression delayed. COPD typically is the result of long-term smoking.

———————

Robert—not his real name—has COPD. When he exerts any physical effort, he almost immediately feels the need to breathe faster. But regardless how fast he breathes in and out—to the point where he is panting—he still feels that he is not getting enough air. At the same time, his pulse rises significantly.

There is a simple explanation for all this.[33]

The air rushing into Robert's lungs has the life-sustaining oxygen needed by the body. That oxygen is absorbed inside the lungs into the bloodstream. It then travels to all parts of the body, where it is taken up by cells and used to burn carbohydrates. The cells expel the resulting carbon dioxide (CO_2) into the blood, which transports it to the lungs. There, CO_2 becomes part of the air to be exhaled.

For this process to work, the lungs must efficiently absorb oxygen from the inhaled air and place it into the bloodstream. With matching efficiency the lungs must pass CO_2 from the bloodstream into the air to be exhaled.

Due to COPD, Robert's lungs have a significantly reduced capacity for these two functions, and he needs to breathe faster.

In terms of models, we may characterize the situation as follows.

A part of the subconscious brain, the *medulla*, contains a model that evaluates the oxygen content in the body's cells.[34]

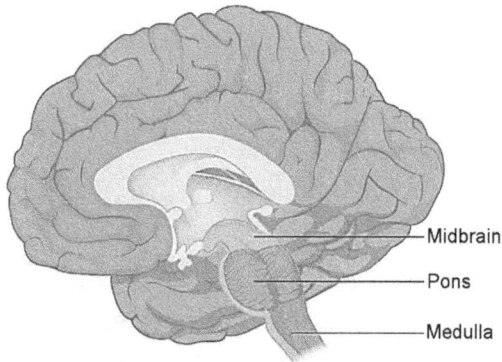

Location of medulla in the brain.[35]

When that content declines for whatever reason, the subconscious model decides that more rapid breathing and increased blood flow are needed.

Robert normally breathes through the nose. But the subconscious model decides that breathing should now be done through the mouth to reduce resistance encountered by the increased volume of air. So without Robert noticing, the breathing switches from nose to mouth.

If the increased breathing and pulse don't have the desired effect, that is, if the oxygen level in the body's cells doesn't go up, the subconscious model outputs to the conscious brain a feeling of breathlessness and ultimately of suffocation.

As that feeling intensifies, the conscious brain tells Robert to reduce physical efforts.

Alternately, Robert may increase oxygen content in the inhaled air with supplemental oxygen.

So the output of the subconscious model leaves Robert with two options: Do less, or use supplemental oxygen.

The decisions made by the subconscious model, and the two resulting options, seem entirely appropriate. But the model actually commits a major blunder.

The error is the subconscious decision to switch from nose-breathing to mouth-breathing. That change is supposed to decrease resistance encountered by the increased air flow. It does so, of course.

But the switch also creates a major problem, as we see next.[36]

Imagine oxygen-rich blood flowing past a cell. The cell absorbs some of the oxygen. Once inside the cell, the oxygen reacts with carbohydrates and creates CO_2. Finally, the cell expels the CO_2 into the blood.

The oxygen in the bloodstream is bound by a chemical bond that must be broken if the oxygen is to enter the cell. How is this accomplished?

For the explanation we must replace the simplistic notion that blood leaving the lungs carries only oxygen, while blood going back to the lungs has only CO_2, with the following, more accurate statement:

Blood always contains a mixture of oxygen and CO_2. It's just that the ratio varies: Oxygen-rich blood leaves the lungs, and CO_2-rich blood returns.

It turns out that the ease with which the cell extracts oxygen from the blood depends crucially on the amount of CO_2 in the blood. That is, the more CO_2 is present, the easier the cell accomplishes this feat.[37]

Thus, efficient extraction of oxygen from the blood doesn't just require plenty of oxygen in the bloodstream, but also a sufficient amount of CO_2.

How can these two goals be achieved?

Well, the oxygen content depends on the amount of oxygen supplied by the lungs.

What about sufficient CO_2 content? Put differently: What action *reduces* CO_2 content in the blood and slows down the transfer of oxygen into the cells?

The answer to the last question is: Rapid breathing depletes CO_2 of the blood.[38]

We now have the following out-of-control situation:

When the oxygen level in the cells drops due to physical activity, the model switches to rapid mouth-breathing. Then the CO_2 level in the blood drops, and the cells absorb less oxygen.

In response, the subconscious model accelerates breathing even more, which further lowers the CO_2 level in the blood, which in turn reduces the oxygen level in the cells even more, and so on.

Eventually, the oxygen level of the cells spirals down to the point where Robert feels he is suffocating.

For recovery, he stops all physical efforts and waits for the breathing to slow down.

We just have seen that mouth-breathing is wrong. But on the surface, the much slower nose-breathing doesn't seem to be an appropriate alternative, since it doesn't push enough oxygen-carrying air into the lungs.

That argument relies on the simplistic notion that the lungs extract *all* oxygen from the inhaled air and replace it by CO_2 to be exhaled.

In reality, the lungs extract only about a quarter of the available oxygen. The rest is exhaled again.[39]

As a result, the slower nose-breathing extracts just a bit less oxygen from the air than the more rapid mouth-breathing. But the increased CO_2 content of the blood more than makes up for that small oxygen reduction, and results in increased absorption of oxygen by the cells.

―――――――――

The conclusion is: Robert must breathe through the nose even though the subconscious model has decided to switch to mouth-breathing.

He can enforce nose-breathing by having the conscious brain override that urge.

A better approach is to systematically replace the subconscious model so that he automatically breathes appropriately.

Doesn't that situation remind you of Chapter 4, where incorrect feelings of fatigue were to be eliminated? Indeed, it is the same case, except that we want to eliminate the urge to mouth-breathe.

The same remedy applies: Robert goes through proactive thoughts such as "Mouth-breathing makes things worse," and "Nose-breathing will help."

He also thinks about details of the oxygen-to-CO_2 exchange and uses a watch to determine how long each inhaled and exhaled breath takes. Finally, he observes how the urge for faster breathing subsides as CO_2 content in the blood increases.

Evidently this is another case of the cognitive behavioral therapy (CBT) of Chapter 3.

Robert engages in these thoughts as he carries out his activities, all the time monitoring that his mouth stays closed and breathing is done solely through the nose.

After a few days of this conscious effort the urge for mouth-breathing subsides, and slow nose-breathing becomes the subconscious choice whether he is at rest or engaged in physical activities.

This also becomes the standard way at night. The evidence is that, when he awakens during the night, his mouth has not become dry as happened before.

He has solved the problem!

———————————

Here are some statistics brought about by the change to nose-breathing. Robert's age is mid 70s. He has smoked for five decades.

- Physical exercise 1: Daily walking.
  - With rapid mouth-breathing:
    Walking 3/8 mile is strenuous. Speed not recorded. Pulse is 94 beats per minute (bpm). Oxygen saturation is 94 percent.
  - With slower nose-breathing:
    On the first day, distance is doubled from 3/8 mile to 3/4 mile, walking at 4 mph. Oxygen saturation climbs from 94 percent to 97 percent. Previous pulse of 94 bpm goes down every day and reaches plateau of 84 bpm within two weeks. After six weeks, distance is increased to 1 mile.
- Physical exercise 2: Walking 50 feet from the house to the mailbox, with a steep elevation drop of six feet. This sounds like a simple task, but isn't easy since he climbs a 12 percent slope back up to the house.
  - With rapid mouth-breathing:
    Robert hyperventilates and becomes exhausted, with pulse rate 90-100 bpm.
  - With slower nose-breathing:
    The time for inhaling and exhaling is the same as at rest, the pulse stays at 80, the oxygen level has increased, and he no longer feels pressured or exhausted.

Do you recognize something else in the statistics for the second item? There is no way that the short trip to the mailbox should cause Robert to hyperventilate and become exhausted. It surely looks like a mild panic attack.

As a remedy, Robert not only has switched to nose-breathing, but for a while used the CBT technique described in Chapter 3 to work against the panic attack. The improvements are then due to both the nose-breathing and the successful fight against the panic attack.

We have seen three examples where replacement of subconscious models improves outcomes: The changes eliminate anxiety attacks, introduce correct response to fatigue, and ease breathing problems. In each case, conscious thoughts insert the new model, as proposed by CBT.

The subconscious model of the next chapter optimizes a basic human activity. As we shall see, the model is easily influenced by conscious thoughts.

6

Motion

You are flying a small plane high up, just below the clouds. The engine purrs, there is plenty of fuel left, the weather is perfect. You enjoy the views: The trucks, houses, bridges, roads, lakes, meadows, and valleys below constitute an extraordinary model-train display.

Suddenly the plane's engine sputters, then stops. Your heart seems to follow suit. Fortunately that's just a mistaken interpretation of the shock you feel.

Contrary to popular opinion, a stopped engine doesn't cause the airplane to crash. Simply put, the wings hold the airplane up in the air, not the engine; the latter is only needed for propulsion.

So with the engine no longer working, your plane still flies just fine, except that you must use gravity and the altitude above ground as the source of propulsion instead of the engine.

For an efficient use of that altitude, you reduce the speed of the plane to the *best-glide speed*: It gives you the maximum distance for each foot of altitude lost. Equivalently, if the engine was running, the best-glide speed would give you the maximum distance for each gallon of fuel used by the engine.[40]

You check if you can glide to the nearest airport using the best-glide speed. If so, you fly to that airport and land there. Otherwise you select an alternate landing site: a large meadow, a road without traffic, or really anywhere you can put the plane down safely.

Once you have figured this out and are en route to the selected landing site, you analyze why the engine stopped, and try a restart. If that effort fails, you land.

The best-glide speed is published in the airplane's pilot manual. Since it is so important for emergencies, you store it in your brain among other crucial information, such as the emergency radio frequency.

The above process sounds perfect, doesn't it? But the computation of best-glide speed and its use are somewhat flawed.

- The best-glide speed is measured by the manufacturer of the aircraft at maximum allowable weight. If the aircraft's weight is below that maximum, the best-glide speed decreases. You can therefore only guess what best-glide speed actually is.

- While you are descending at best-glide speed, the winds aloft will affect your *ground speed*, the speed across the terrain. A headwind will slow you down, and a tailwind will accelerate you. In the first case, you should fly faster than the best-glide speed, and in the second case, slower.

- While you are descending, you initially may have a headwind, but closer to the ground a tailwind. So initially you should fly faster, and later slower than best-glide speed.

Given these complications, you cannot determine the optimal speed while coping with a stopped engine. Even if you knew the total weight of the plane and the wind speeds at the various altitudes below you, computing the optimal speeds as you descend would be a formidable mathematical optimization problem.

So instead you fly the published best-glide speed. You estimate whether you can reach the nearest airport by correcting that speed with a guess of the head- or tailwind. If your plane is well-equipped, software carries out these computations.

The process is a bit crude, wouldn't you say? In contrast, let's see how evolution has dealt with the problem of efficient human motion.

_____

Mankind began to domesticate large animals for transportation—for example, the donkey, camel, and horse—about 10,000 years ago.[41] Until then, walking and running were the only ways to move on land.

Compared with 2-3 million years of human development,[42] 10,000 years of transportation using animals are a blip in time. Thus, evolution had plenty of time to refine walking and running.

We focus here on walking. One may describe that process as the movement of an inverted pendulum where the body vaults over a stiff leg during each step.[43]

Suppose we want to hike a given distance, say 1 mile, at a speed that minimizes the total energy used by the body for the entire trip. We completely disregard here how long this will take. Let's call that speed the _best-walk speed_. It is the analog of the best-glide speed for airplanes.

The best-walk speed depends on numerous factors, among them height, weight, age, load, terrain, surface, altitude, and fitness.[44]

It may seem rather difficult to determine the best-walk speed: To take all factors into account, we must measure energy consumption for various walking speeds, where each time we cover the given mile. The speed with minimum energy consumption is then the best-walk speed.

The energy consumption during those trials can be established in various ways. The most frequently used method measures oxygen consumption with a face mask where the difference between the oxygen quantities entering and leaving the lungs is recorded.[45]

_____

There is a much simpler way to obtain the best-walk speed: We walk a few minutes at the speed that feels most comfortable. Toward the end of that short period, we are walking at the best-walk speed! How does the brain achieve this miraculous feat?

While we were growing up, the subconscious brain learned efficient walking during thousands of hours of practice. The subconscious brain has a model that constantly tunes and updates that established pattern. Within minutes, the brain achieves minimum energy consumption per mile of walking for current circumstances.[46]

When the speed is forced below or above that optimal value, the model outputs to the conscious brain a feeling of annoyance: The low case produces impatience, and the high case discomfort. At the optimal value there is a feeling of pleasant achievement.

We can readily change our walking speed from best-walk to slower or faster values while ignoring the feeling of impatience or discomfort. Thus we can easily override the model of the subconscious brain controlling walking speed.

For example, we may want to arrive sooner due to some commitment, and thus walk at a brisk rate. Or we may decide to slow down to a crawl, for example when we visit an art gallery. In either case, the goal we are pursuing with the changed speed provides a pleasant feeling that overrides the negative emotion arising from the nonoptimal speed.

We turn to subconscious models handling the body at rest. At times, they produce quite undesirable output. The next chapter shows how we may change these models and thus improve our lives.

7

Rest

Life in the 21st century is filled with stress.

- During the day, we almost continuously soak up information from the Internet.
- Social media and email ping for our attention around the clock.
- Fast-paced video games tax mental and physical limits.[47]
- The work environment demands that we react instantaneously to any and all demands.

Giving our brain and body a rest has become difficult. When we are supposedly unwinding, the conscious brain keeps on thinking and worrying, and the subconscious brain holds the body in a state of tension and readiness for the next demand.

In the olden days—here meant to be the time before computers took over, in the 1990s—there was less bustle and pressure. So when one rested, the conscious brain slowed down the flood of thoughts to a trickle, and the subconscious brain allowed the body to let go, so to speak.

We may return to those days with conscious actions. For example:

- At certain times, we turn off computers, tablets, and phones, and fill those periods with relaxing activities such as walking in the park, reading a book, listening to music, gardening, painting, baking, or dancing. These old-fashioned activities make the brain slow down: It no longer rushes along and pushes the body into a tense performance.

- When we are on vacation, we impose a *news fast* where we do not read newspapers, watch TV, or surf the Internet for news.[48]

The proposed changes may look like a dull life. The negative feeling comes from subconscious models that produce boredom when we slow down, and generate happiness only when brain and body are pushed to their limits.

How can we change those models?

Reading books about the proposed changes may move us in the right direction: We agree with the helpful suggestions and implement them for a few days or weeks. But thereafter, we slide back into our old habits.

Indeed, the lure of the fast-paced life style of the 21st century is so strong that our old subconscious models soon are active again and rule our lives.

Instead of a one-time makeover, we may aim for a change in our lives where we reinforce the desired subconscious models *every day*. This is very different from the typical goal of psychotherapy interventions—for example as described in Chapter 3—where we change subconscious models permanently for the better.

The reinforcement is attained by daily *meditation*. We believe that meditation isn't just useful for this purpose, but essential for the pursuit of a simpler, more relaxed, and more enjoyable life.

You may pause for a moment and look up the "Meditation" entry of the Wikipedia.[49] It is an amazing assembly of diverse human thought on various ways to change our lives.

We shall not attempt even a cursory overview of the possibilities. Instead, we focus on one particular approach that is easy to learn and doesn't require much time every day, yet gives us a new sub-conscious model that allows us to live contentedly without pushing mind and body to their limits.

The technique is *mindfulness*.[50]

It creates a value system that rejects proposals such as "Work hard and play harder," and instead allows us to experience the world in all its richness without the artificial machinery of electronics.

After a while, our perception of the world has shifted: We enjoy living in the present without constant thoughts of past and future and feel more connected with nature and people.

Part of mindfulness is a *body-scan technique* where we lie down, relax, let breathing and pulse stabilize, and gradually survey the body. We experience what each body part feels at the moment. The effect is that we not only learn to listen to our body, but also become sensitive to the world around us.

The body-scan technique requires determined training[51] in which we follow the steps of the program every day.

In the language of subconscious models, we gradually install a new way of viewing the world. It overlays the old electronics-driven subconscious model.

At the end of each session we congratulate ourselves for having done the meditation and end with a strong, positive feeling.

After a few days, we begin to feel more connected with the world. When we take a walk in the park, watching and admiring the beauty of trees and lakes and birds, we look at other walkers staring at a phone with mild amazement. Why don't they see the beauty around them instead of focusing on a tiny glass plate with miniature pictures and text?

Later, we learn to enjoy the meditation process itself: It is a nice experience, independent of the fact that it produces a richer and more relaxed life.

Later still, we realize that our entire lives can be viewed as a meditative process. At that point, we have become a different person.

Mindfulness meditation prepares us well for a change of other detrimental subconscious models. An example is anger management.

During a meeting at the office, somebody launches a heated attack on us. At first we feel annoyed, and then angry.

What should we do? In childhood we learned to fight back, using our anger as a driving force. If we do so, our anger will rise as we respond. Then there will be counterarguments and even angrier reactions by us.

Not good, is it?

Buddhism has an alternate approach.[52] It views our rising anger as both the problem and the opportunity where we learn and grow. That is, the attacking opponent is not a foe, but makes it possible for us to practice effective anger management.

These thoughts disconnect the anger from us and turn it into an object we want to manage.

When the discussion has run its course with us calm and relaxed throughout, we tell ourselves, "Hey, isn't it great how we handled this situation and became better at handling anger."

The subconscious model triggering that reaction can be learned just by reading about it. We reinforce the model by regularly thinking about it.

Isn't it wonderful that Buddhism has given us this tool?

We have seen several examples where changing a subconscious model is difficult. The next chapter shows that these weren't typical cases. Indeed, many subconscious models are easily modified.

8

Training

We exercise daily, or should do so, to enhance endurance, muscle tone, and oxygen saturation; in short, to improve physical performance of our body.

Every day we also update many subconscious models.

- When a piano virtuoso practices nine hours every day, she sharpens subconscious models that operate the fingers of her hands with lightning speed. During the concert, she doesn't consciously trigger each of these movements, but allows the models to operate the muscles. Instead, she focuses on the mood and tone and flow of the music.

  She must update these subconscious models daily since otherwise the precision of muscle control deteriorates.

- When a lab technician operates some equipment for the first time, conscious models control the steps. But repetition of the steps inserts parts of the process into subconscious models.

- When we write or type text, muscle movement is mostly controlled by subconscious models. At the same time, the activity updates the precision of the models. We become aware of this when we haven't used a particular skill for a while: The task seems difficult, and our performance is error-prone.

- Coming back from a vacation, our home seems strange: Things are not quite as we remember them. The same goes for driving a car or operating complicated equipment: When not done for a while, we gingerly and consciously keep track of the steps until the subconscious models have become fully functional again.

Various subconscious models control the body: for example, blood pressure, pulse, and breathing. The above training cases are different: When we repeatedly carry out conscious operations, we train subconscious models to carry out these operations partially or fully without conscious supervision. For example, the speed with which a piano virtuoso plays the third piano concerto by Rachmaninoff— a devilishly difficult composition[53]—is such that conscious models couldn't possibly supervise the playing of each note.

The playing of the notes is not evidence of *decisions* by subconscious models, since the term "decision" is reserved for conscious models. But it is appropriate that we call the playing an *action* of subconscious models. Of course, there is the conscious supervision of mood, tone, and flow of the music.

Similarly there are actions of subconscious models when we drive a car, operate lab equipment, and so on. In each case, the conscious brain may be partially or fully or not at all aware of the action.

We now have three ways in which subconscious models produce output: They generate feelings, trigger thoughts for conscious models, and carry out actions.

We move to the second part of the book, where we examine conscious models instead of subconscious ones.

During the past 130 years, conscious models have produced a world of magic and wonder. For example, in 1891 Otto Lilienthal[54] a-chieved the first heavier-than-air flight. In 1903, Orville and Wilbur Wright[55] accomplished the first powered heavier-than-air flight. In 1969, man flew to the moon and returned safely. Since 2000, the International Space Station orbits the earth.

We could go on, recounting other miraculous feats in the sciences and industry. We could discuss the Internet, the Global Positioning System, today's powerful computers, and the worldwide communication network.

These developments have already been described in great detail elsewhere. Just pick a topic and go to the Wikipedia for references. In fact, the Wikipedia itself is one of the modern wonders: a living encyclopedia that truly covers everything and is always up-to-date.

Instead of repeating some of these success stories, we look at a much-less covered dark side of conscious models: They are declared to be useful and benign, yet produce harm, sometimes on a stunning scale.

The next four chapters cover such deceptive models. The first case occurs in medicine.

Part II

Deceptive Conscious Models

9
Medicalization

Medicine has created an admirable model for treating diseases. It prescribes analysis of the complaint, physical examination, lab tests, diagnosis, selection of treatment, verification of effectiveness, follow-up process, and prevention.

The model has worked very well. Some diseases have been completely eliminated, while others are mostly under control. There are few notable exceptions—for example, autoimmune diseases, certain types of cancer, and virus infections.

The success story has a dark side. At times, the medical model has been applied to human troubles that were not diseases, with disastrous if not horrific outcomes.

Since 1970, such abuse of the medical model is known as *medicalization*.[56] The fact that the word was coined at that time doesn't mean that the malpractice didn't happen before.

Here is a horrific example from the 19th century. In 1851, the physician Samuel A. Cartwright advanced the hypothesis that enslaved Africans trying to flee captivity suffered from the disease *Drapetomania*. The medical cure for the disease was removal of both big toes, thus making running a physical impossibility.[57]

If an enslaved African was "sulky and dissatisfied without cause," supposedly a symptom that flight was imminent, Cartwright prescribed "whipping the devil out of them" as a "preventative measure."

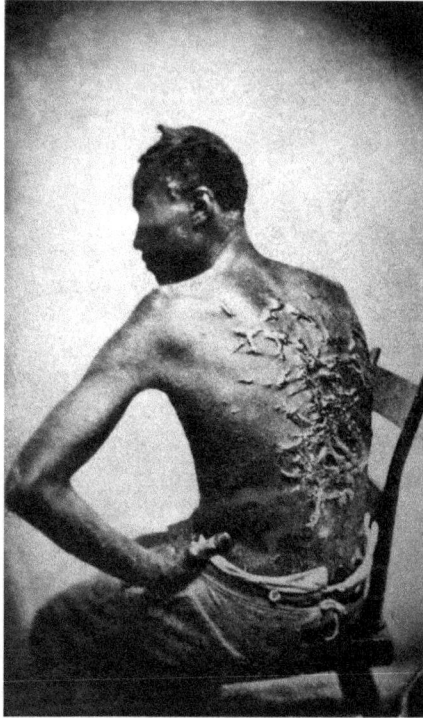

Lesions on the back of an enslaved African from Mississippi.[58]

Cartwright identified a second disease, *Dysaesthesia aethiopica*, that supposedly caused laziness. He declared insensitivity of the skin to be a symptom, to be eliminated as follows: Wash the skin, anoint it with oil, slap in the oil with a broad leather strap, and impose hard work in sunshine.[59]

Fortunately, such cruel medicalization no longer takes place.

―――――――――

Fifty percent of the people of the Western World have been medicalized since antiquity.

You guessed right: the women.

Essentially, their sex drive and the natural processes of their reproductive system were turned into medical problems requiring intervention. Three examples:

- Universal female disease: *hysteria*
- Planned childbirth: *cesarean section*
- Menopause as a disease: *hormone replacement therapy*

Let's examine these cases. The tale of the presumed mental disorder *hysteria* began in ancient Egypt and continued through the ages. Hysteria supposedly was indicated by a large array of symptoms that varied during hundreds of years:

Anxiety, shortness of breath, fainting, nervousness, insomnia, fluid retention, heaviness in the abdomen, loss of appetite or sex drive, but also sexually forward behavior, to name a few.[60]

At the core, it is the reaction of a woman who must repress her sexuality and liveliness to conform to standards of a male-dominated society.

The disease originally was attributed to a wandering uterus. The name *hysteria* reflects this, since it is derived from the Greek word "hystera" for womb.

———————

Cesarean section is surgery to deliver babies. It should only be done when vaginal delivery would put baby or mother at risk. Under that criterion, approximately 10–15 percent of deliveries should require that surgery.

In the U.S., the rate for 2011 was far higher at 33 percent. This indicates that there were a considerable number of cesarean sections for low-risk pregnancies. For those cases, the surgery is more prone to create complications than natural births.

Indeed, adverse outcomes for low-risk pregnancies in 2011 in the U.S. had an incidence of 8.6 percent for vaginal deliveries and 9.2 percent for cesarean section deliveries. That increase was accompanied by a tripling of maternal mortality: from 3.6 deaths of the mother per 100,000 deliveries to 13.3 per 100,000.[61]

In Northern Europe, the rate of cesarean sections is 14 percent. In the Rome region of Italy, it is as high as 85 percent in some private clinics.[62]

So why do Northern European countries have such a low rate compared to the U.S., and some private clinics in Rome have an astronomically high rate?

The attending doctors make these choices, apparently driven by the convenience and speed of a scheduled delivery. The mother-to-be readily agrees since it promises a painless birth.

But in the case of a low-risk pregnancy, would she still make that request if told by her doctor, "That choice makes it three times more likely that you will die"? Seems unlikely, doesn't it?

Isn't it the doctor's duty to insist on the choice that is best for mother and baby?

Hormone replacement therapy (*HRT*) is justified if the natural production of hormones has ceased in a woman due to removal of the ovaries.

But the term takes on an entirely different meaning when menstruation stops in a healthy woman in menopause and hormone levels change over subsequent years as part of the natural aging process.

The term "Replacement Therapy" then implies that the stop of menstruation leads to a hormone deficiency that must be continuously combatted as a woman gets older.

HRT started in the 1960s, reaching maximum popularity in the 1990s. The first large-scale clinical trial of HRT by the *Women's Health Initiative* was launched in the late 1990s, more than 30 years after the introduction of HRT.

Isn't that astonishing? For three decades millions of women took powerful drugs without any testing of the long-term impact.

By 2006, the trial proved that HRT for healthy women increased rates of breast cancer, stroke, lung clots, endometrial cancer, gallbladder disease, urinary incontinence, and dementia. Even when hormone treatment was discontinued, there was an increased risk of breast cancer.[63]

A few factors on the positive side—such as reduced rates of hip fractures and diabetes—did not balance out the long list of major negative results. Indeed, parts of the study were halted ahead of schedule because of convincing negative findings.[64]

The reaction of the pharmaceutical industry to the negative results were new hormone formulations.

The history of HRT leads to the following conclusions:

- The iterative process of negative results of HRT for healthy women on one hand and of new hormone formulations that supposedly avoid those problems on the other hand, can be carried forward indefinitely, with healthy women always playing the role of test subjects.

- Underlying that process is the notion that hormonal changes starting with menopause in a healthy woman are a disease that requires treatment, and that one just needs to find out what that treatment consists of.

- By planting that idea into the heads of physicians, the pharmaceutical industry effectively has turned healthy women into a huge source of never-ending revenue.

During the last 100 years science has produced many drugs that fight diseases once thought incurable. Many results seem almost miraculous, such as the obliteration of infections by antibiotics and the prevention of diseases by vaccines.

In recent decades, the preventative use of drugs has been hugely expanded.

People one might call *potentially healthy* are urged to take drugs on a regular schedule. The term "potentially healthy" means that health is compromised by poor lifestyle choices.

For example, many potentially healthy people are overweight and have high blood pressure, caused by lack of exercise and consumption of too much food loaded with salt, sugar, and fat.

The pharmaceutical industry, Pharma for short, has taken the lead in this medicalization process, by creating drugs and testing whether potentially healthy people live longer when they take them on a daily basis.

For successful marketing, Pharma has to convince doctors to prescribe the preventative drugs. Pharma does this with various marketing strategies, where for example the visit of a marketing representative becomes a fun event in the doctor's office[65] and where attendance of conferences by the doctor in exotic locations is financially supported.

Similar arguments apply to the treatment of certain illnesses. There may be an alternative involving exercise and a special diet that actually is more effective than drugs. Examples are the control of certain cases of diabetes by a prudent diet and the management of back pain by exercise.

Why do people prefer the consumption of drugs to a change of lifestyle? There is a superficial answer: The choice is due to the ease of taking some pills versus the complex task of changing a lifestyle.

A better answer is that the choice depends upon the model we have in our brain about the goals of our lives and how much effort we are willing to expend to get there.

That model resides in the subconscious brain. Changing it is a difficult task. But it can be done with insight provided by the relevant literature,[66] just as we saw in Part I.

The next chapter discusses deceptive conscious models of economics.

10

Misevaluation

There are widely accepted conscious models that aren't just deceptive, but constitute huge blunders.

Sometimes we are lucky and avoid the use of such a model despite its popularity. Here is an example.

The *Club of Rome* is an informal organization that attempts to understand global systems and promote policy initiatives and actions. A group of thirty individuals—scientists, educators, economists, humanists, industrialists, and public servants—started the program in 1968.[67]

In 1972, the Club of Rome published the book *Limits to Growth*.[68] It used mathematical models to predict future economic developments up to the year 2000, about 30 years ahead.

The main conclusion was that economic growth could not continue indefinitely due to excessive consumption of resources.

The book caused a huge stir. Eventually 30 million copies were sold in more than 30 languages, making it the biggest best-seller of all time about the environment.[69]

An implicit conclusion of the book was the following: The major industrialized nations of the world needed to scale back production so that there would be enough resources left for the economic development of poor countries.

Fortunately for the world, the industrialized countries did not adopt that proposal. Instead, a tsunami of technological innovation during the 30-year prediction period and beyond lifted countries out of poverty in steady progression. At the same time, the industrialized countries improved their standard of living. The universal improvement process continues unabated today.[70]

An example: In 1968, the gross national product per person of the U.S. was larger than that of China by a factor 44. In *Limits to Growth*, it was predicted that under ongoing policies and practices that factor would *grow* to 110 by the year 2000.[71] In reality, the factor *dropped* from 44 to 38.[72] The reduction marked the beginning of an astonishing development in China that by 2019 had slashed the factor to 6.[73]

There is another measure of economic progress in China. In 1997, 42 percent of its population lived in extreme poverty. By 2017—just 20 years later—that share had dropped to an amazingly small 0.7 percent.[74]

Contrary to the predictions of *Limits to Growth*, the economic power of industrialized nations lifted many countries out of poverty. This would not have happened if the industrialized countries had self-limited their economic development. Indeed, the world was lucky that the conclusions of the widely accepted, supposedly convincing, but nevertheless erroneous models of *Limits to Growth* were disregarded.

––––––––––

Next is an erroneous model that is the opposite of the above case: For years it has been known that the model is badly flawed, yet it is widely used to make decisions. We begin with a story.

In the 1960s, the Union Electric Company was a large electric utility headquartered in St. Louis, Missouri.[75] A stockholder asked the chief executive officer (CEO) of the company during the annual general meeting why he, as CEO, didn't own any stock of the company. Was he worried that the stock would not perform well?[76]

The CEO hadn't forgone stock ownership because he thought the stock wouldn't be a good investment. He felt that, by owning shares, he would favor choices producing short-term increases in stock price when he was supposed to focus on the most important factors: reliable supply of electricity for customers, long-term success of the company, proper treatment of employees including health care and pension plans, and regard for environmental concerns.

The day after the stockholder meeting, the CEO bought a significant number of shares so he wouldn't have to listen to such unfair criticism.

Starting in the second part of the 20th century, the key goal of the CEOs of large companies became maximization of *market capitalization*, which is the total worth of the issued shares and thus represents the total value of the company for stockholders. It is computed by multiplying the number of outstanding shares with the current market price per share.

Since the number of outstanding shares generally doesn't vary much, market capitalization is maximized by driving up the stock price as much as possible.[77]

The CEO of Union Electric Company of the 1960s looks like an old-fashioned fool to these executives. In their opinion, he paid too much attention to employees, bond holders, customers, the community, and the state. Instead, he should have solely focused on stock price.

Why would a CEO go for this single goal, instead of the much more differentiated aims of the CEO of Union Electric?

The answer: Since the 1970s, the system establishing executive pay has been structured such that compensation of the CEO rises in lockstep with stock price. Hence maximization of stock price maximizes the CEO's compensation.

How did we get to that rule?

The story of executive compensation systems over the past 120 years is too complex to permit a simple answer.[78] Generally speaking, there have been several cycles where huge executive compensation was hidden from view but eventually was discovered, and where regulatory agencies and the U.S. Congress stepped in to curb the excess. The resulting laws triggered new ways to hide exorbitant compensation and started the next cycle.

The history of these cycles has been likened to the story of the Dutch boy who plugs a leak in a dike with a finger. But contrary to the story—where the boy saves the country—he then discovers more leaks.[79]

Let's bring the story to its logical conclusion. The Dutch boy runs out of fingers to plug the leaks, the dike fails, and the country is flooded. That's the current situation for executive compensation.

So far we have only mentioned compensation of the CEO. But the pay of executives immediately below the CEO is also based on stock price. Going further down in the organization, yearly bonuses depend on stock price as well.

Let's see how maximization of stock price has influenced upper-level decisions.

- Top management decides to borrow an excessive amount of funds to accelerate the growth of the company. The decision boosts short-term profit and thus stock price. But it also increases the odds that the company eventually will go bankrupt.[80]

  This is a high-stakes gamble with enormous personal gain for top management and potentially huge losses for the other stake holders: the employees, bond holder, customers, the community, and the state.

  The community and state are affected since tax money will be used to pay the social costs of bankruptcy.

- When the company buys back its own stock, the stock price goes up. Hence this is a favorite move by CEOs.

  It is a poor decision since it reduces cash reserves that might become essential for long-term survival of the company.

- Pension plans are a substantial liability for companies and can become a major drain on profit.

  To reduce expenses, CEOs replace pension plans with defined-contribution plans in which companies contribute regularly to employee-owned retirement accounts.

  This increases stock price. But it doesn't work well for the employees since they often are poor managers of their retirement accounts. They put off saving, don't know how to invest, and pay large fees for investment management. The combined effect is a vast retirement crisis.[81]

- Whenever possible, top management moves manufacturing to countries with lower labor costs to boost profits and, hence, stock price. But the U.S. employees lose their jobs and often must accept lower-level jobs at much reduced pay.

  We should mention that there is also a significant upside: The movement of manufacturing jobs has lifted poor countries out of poverty. It's just that the change should be done gradually in the U.S. and with significant assistance for retraining.

- CEOs carry out actions that boost the stock price short-term—say for the next two or three years—but later hurt company operations. An example is a temporary reduction of research and development activity. It lowers operating costs, thus increasing stock price. But the decision may imperil long-term survival of the company.

Traditional manufacturing relied on the following model, borrowed from agriculture:

The employees manufacture the product. The company sells the product at a profit. The customers buy the product, use it, and return for more. In effect, the purchases of the customers supply the wages paid to the employees. Modern high-tech industry operates differently:

The employees design a product and build automated equipment for its manufacture. The company makes copies of the product at low cost and sells them to customers, who use them and come back for more. Effectively, the employees below senior management get mostly paid for the design of the product and the construction of the automated equipment, but obtain little income from subsequent sales.

An extreme case is software production: Once a computer program has been created, it can be reproduced at zero cost.

CEOs of high-tech industries exploit these facts to maximize stock price as follows:

- Salaries of employees are increased once a year just to account for inflation. There is also a yearly bonus, which below senior management is computed as a percentage of salary. Since the bonus doesn't affect base pay, the sum total of salary plus bonus never changes when adjusted for inflation, apart from the ups and downs of the bonus.

- In contrast, increasing sales in a growing economy boost profits and thus stock price. Hence, compensation of top management climbs, sometimes to stratospheric levels.

Let's look at the overall effects maximization of stock price has wrought:

- The ratio of executive pay divided by average worker pay[82] grew from 21-to-1 in 1965 to a much larger 61-to-1 in 1989, and to an incredible 320-to-1 in 2019, an overall growth of the ratio from 1965 to 2019 by a factor of 15. It is an obscene valuation of the executives' contribution.

- The elimination of pension plans and the poor management of retirement funds by employees combine to create a troubled financial future when the employees retire.

- Since most stocks traditionally are held by the wealthiest percentage of the population, the almost fantastic growth of stock prices of recent decades has produced the following result:[83]

  During the period from 2000–2020, the portion of the U.S. wealth owned by the top 1 percent of the population grew from an already large 25 percent to an astonishing 31 percent. Thus, in 2020 roughly one third of the entire wealth of the U.S. was in the hands of 1 percent of the population.

- The lower 50 percent of the population has done very poorly:[84]

  In 2000, the wealth of the top 50 percent of the population was larger than that of the bottom 50 percent by a factor 29. By 2020, just 20 years later, that factor had almost doubled to 53. In fact, the wealth of the bottom 50 percent had hardly changed over that period.

These effects will worsen as automation and technology take over.

Clearly a different model for executive compensation is needed. The main goal should be a fairer division of the economic pie of the U.S.

The next chapter deals with a very different kind of exploitation.

11

An Alternate World

A discussion of opposite viewpoints requires a large measure of goodwill on all sides. Indeed, it is easy to sabotage the process by throwing in irrelevant arguments. Such detraction assures that the discussion is going nowhere.

In recent years we have seen a destructive approach to discussions that goes far beyond the disruption by irrelevant arguments: It threatens the foundation of the U.S. democracy. This chapter analyzes the process.

Model-dependent realism postulates that we consider the output of our conscious models to be facts. So when we talk about the movement of the earth around the sun, we envision an elliptical path. When we see a half-moon, we declare that this is the first or last quarter of the moon's phases.

When scientists propose a model, we demand verification by tests. This is particularly important when a new model claims to replace an old one.

For example, Einstein's model of the universe as formulated in the theory of relativity was in conflict with Newton's model. Careful verification justified that replacement.[85]

Suppose we declared today that the theory of relativity is bunk. What would happen? Scientists would speak up, point out that the model of the universe produced by the theory has been repeatedly tested and found to be valid.

In the past, that would have settled the issue. Indeed, if we had persisted in our claim, we would have been ignored and consigned to the bin of lunatics.

But now there is a new way to carry on with our idiotic claim:

- We do not just make the erroneous claim in a technical publication or newspaper article, but publish it on numerous websites and spread it on social media.

- When alarmed scientists respond, we declare their statements to be false, without any proof. We add spurious claims about the scientists. For example, we claim them to be in collusion. It doesn't matter that our statements are wacky.

- The scientists respond again, this time on a broader front using national newspapers, radio, and TV.

- Once more we declare these statements to be false without proof. More importantly, we attack the newspapers, radio stations, and TV programs that have stated their rebuttal. We declare that these media outlets always put forth bogus information. Collectively, we call them the "fake news" media.

- We find a national TV channel that supports our idiocy. That channel continuously broadcasts our claim: The scientists are making false claims and are in collusion. The other media publishing the claims are "fake news" media.

- A significant percentage of the population begins to believe our claim that the theory of relativity is wrong. The TV channel favoring us becomes their main source of information. They believe that the remaining media only supply "fake news."

- They become radicalized and demand that the scientist defending the theory of relativity be fired from their positions. During rallies, they shout "Fire them all."

- At this point, we have created a group of people who trust only information coming from one TV channel and reject all other information as fake news. They will believe almost anything we claim on that TV channel and in related messages on the social media.

Sounds crazy, doesn't it?

Well, the process didn't occur for the theory of relativity. But it did happen for another scientific result during the COVID-19 pandemic of 2020–21.

Scientists had computed that the pandemic could be reasonably well-contained in the U.S. until a cure or vaccine was found, if people observed three simple rules: Always wear breathing masks outside the home, wash hands frequently, and avoid meeting in groups.

Due to the above disruptive process, a large percentage of the U.S. population thought that the scientists advocating these rules were fools. These folks ignored the rules and eventually caused a massive number of deaths. A single statistic bears this out: By January 2021, 20 percent of the world-wide deaths of the pandemic had occurred in the U.S., yet the U.S. has only 4 percent of the world's population.

Let's use this method for truly devious goals. Instead of declaring that the theory of relativity is bunk, we now say, for example, that some politicians are in collusion with a foreign country so that the U.S. loses its leading position in the world, or that they advocate the conversion of the U.S. to a socialist country. There isn't a shred of evidence for this, but we make these claims anyway.

When we are done, the people accepting our claims have inserted into their brain conscious models that not only accept our erroneous claims, but also reject any counter information published by the regular media as "fake news."

In some sense, these persons are living in an *alternate world*.

During the years 2016–20, the president of the U.S. managed to pull off construction of an alternate world on a national scale. He convinced 35-40 percent of the U.S. voting population that his world was the true one. That part of the population resided in just one of the two major parties. Effectively, the president controlled more than 70 percent of that party.

As a result, any party member disagreeing with the president couldn't possibly win in election primaries and thus couldn't become a candidate for the general election.

So by 2020 the president had total control of the party, and the important separation of the legislative and executive branches of government no longer existed.

The party had the majority in the U.S. senate. That body of Congress affirms all judicial appointments at the federal level. Since the president proposes these appointments, the president effectively made all judicial appointments: He appointed 234 of 792 active federal judges.[86] This included three judges of the Supreme Court, one third of the total. Thus, the president single-handedly reshaped the federal judiciary.

In essence, the president was not only the top manager of the executive branch, but had an overwhelming influence on the legislative and judiciary branches of government.

In January 2021, thousands of believers in the president's alternate world carried out a violent attack against the U.S. Congress in Washington, DC. The president had lost the election in 2020, but falsely declared that his election was stolen from him by massive voter fraud.

A number of judicial courts determined the claim had no basis in fact. Nevertheless, the president persisted in his false claim and incited a mob of his believers to storm the Capitol. The goal was to disrupt certification of the duly elected next president.

The mob breached police perimeters, entered the Capitol, vandalized and looted for several hours, and almost managed to capture lawmakers. Five people died and more than 140 were injured.[87]

A disastrous development for U.S. democracy, wouldn't you say?

What's being done to prevent the construction of alternate worlds?

* Significant efforts attempt to contain the publication and propagation of false information across social media. Huge expenditures are being made to weed out misleading information. This includes the development of software that identifies erroneous information with the reliability of a human expert.

* There are numerous websites around the globe that fact-check information.[88]

* Each month, Wikipedia's 100+ million pages are searched in 90+ billion requests. At the same time, 20+ million edits help to ensure correctness.[89] If you were to go on any one of these pages and insert wrong information, within minutes somebody around the globe would catch the error and delete it.

Time will tell whether these efforts suffice. We are inclined to guess that additional checking is needed to combat the construction of alternate worlds.

The conscious models discussed so far concern the world. In contrast, the next chapter examines deceptive conscious models that cover not just the world but also beliefs.

12

Expanded Worlds

More than 30,000 years ago, the human mind began to imagine certain forces that controlled events such as weather, diseases, or death.[90]

These forces were later replaced by gods.

In a final step, the gods were connected with historical persons and events, and thus became seamlessly joined with the world.

We call the results of that union *expanded worlds*. Today's religions make various claims about them. In the spirit of this book, the religions offer *models* of expanded worlds.

We discuss here deceptive aspects of the models of the world's most popular religions: Christianity and Islam. Their believers make up more than half of the world's population.

We should emphasize that we don't intend to criticize the *beliefs* of either religion. It would be utterly arrogant to do so. But we do point out parts of the models that are used to control people.

———————

The natural sciences have produced an extraordinary body of theory about the origin and development of the universe:

- The physical sciences have put forth detailed theories of the start of the universe in the Big Bang and the subsequent development of atoms and molecules, and their aggregation to planets, stars, and galaxies.

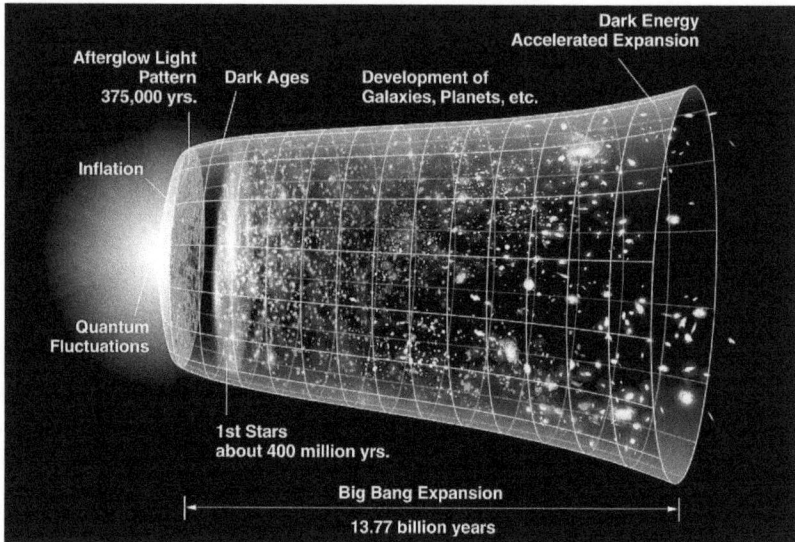

Big Bang model of time and space.[91]

- The life sciences have shed profound insight into the start of the fabric of life and its progression over eons.

If we want to learn about any of this, we can get detailed information from Internet queries. So why is there any need to enlarge the models of our world with the claims of religions?

The question is all the more baffling since early versions of the religions considered here included erroneous claims.

An example is the original assertion of Christianity that the earth was the center of the universe. It was based on a perceived central importance of the earth and its inhabitants.

It now clashes with firmly established models of astrophysics in which the earth is a small part of a minor solar system positioned at the fringe of a galaxy in a universe with more than two trillion galaxies.[92]

In response to such insight of the natural sciences, proponents of the religions considered here eliminated or reinterpreted erroneous claims to avoid embarrassing failures. Today's versions only assert facts that the natural sciences cannot disprove.[93]

One of the untestable assertions says that each person has an eternal soul. Why would a religion make that claim?

We see daily that some people die, and rationally know that sooner or later we will die, too. We try to mitigate that depressing thought: We acknowledge that death happens to other people, but downplay that this applies to us.[94]

The religions considered here promise salvation: If we behave properly on earth and don't engage in bad activities, the eternal soul will leave the body upon death and move to a heaven of eternal bliss.

We usually are sceptical of far-out promises. How is it then possible that billions of people on earth believe in an eternally blissful life?

In an attempt to explain this, we first summarize the steps that create, propagate, and maintain the religions considered here.

- Building upon the claim of an eternally existing god, one person founds the religion, promotes it, and attracts disciples believing in the validity of the religion.

  After the founder has died, his disciples declare that he has achieved eternal bliss and is a prophet or directly related to the god.

  This process seamlessly connects the god with the world.

- The disciples or their successors write down key parts of the religion. This assures consistent propagation.

The religion demands efforts to expand the circle of believers.

- Some believers become missionaries who convince people to accept the religion.
- Believers are pressured to raise their children so they adopt the religion as well. The requirement is based on the correct assessment that, up to puberty, children largely emulate the behavior of their parents. Effectively, the children cannot help but believe in the religion.

  Part of the education are rites that permanently commit the children to the religion.

The religion has rules that prevent believers from leaving:

- Believers are urged not to question validity of the religion and are advised against reading critical material.
- There is a severe penalty for any believer who abandons the religion. This takes various forms, from eliminating all reminders of that person to acting as if the person had died.

  Regardless of the case, hell is claimed to be the final destination for the soul of the former believer.

Subconscious models are extensively employed to support belief in the religions:

- Complex rites create subconscious models that convert the aroma of incense, sound of bells, public calls for prayers, the prayers themselves, and singing of hymns to positive feelings about the religion.
- Elaborate churches and mosques create a similar effect: During each visit, subconscious thinking produces feelings of wonder and awe that validate the religion.

Over time, leaders of the religions introduce rules and regulations that control the behavior of believers.

Some of the rules are based on the idea "This is a good rule for believers to follow," such as helping people in need. Other rules are arbitrary and constitute psychological manipulation and exploitation. Examples:

- Certain actions are declared to be minor or major sins. The characterizations include not only cases anybody would consider reprehensible, such as theft, but also breaking of artificial rules, such as not attending a mandated rite.

  Given the complexity of the demands, it is impossible to meet them all. As a result, a believer always feels guilty for failing to measure up.

- The believer is pressured to carry out demanding tasks like pilgrimages. As long as the tasks are put off, the believer feels guilt and obligation. Once completed, they turn into a significant investment, strengthening the belief in the religion.

- There is a constant threat that the move to heaven may not happen. The soul is then condemned to eternal suffering in hell. The decision depends on the extent to which the believer has measured up.

Given these conditions and potential outcomes, the believer is always under pressure to obey the rules.

We have seen that the religious models of Christianity and Islam promise us an eternal life in contentment and peace. But they also threaten us with eternal damnation if we don't follow rules that are almost impossible to satisfy.

Here are two ways to break out of this web of promises and threats.

We retain the belief in an eternal soul and the core claims of the religion, but reject the threats. We then say that we believe in the existence of a god who is kind and supportive, and reject the notion that our god is punitive.

People making that choice sometimes say that they believe in a god, but are not interested in organized religion.

Alternately, we may adopt a more radical solution consisting of two steps.

In the first step, we abandon the notion of an eternal soul. With that decision, we discard the promise that we can live forever in heaven—a frightening loss. But we also eliminate the notion of hell and other punishments—a welcome liberation. We should mention that this choice is made rarely, as is evident from the fact that just 15 percent of the world's population identifies itself as secular, nonreligious, agnostic, or atheist.[95]

The elimination of the soul and heaven and hell is accompanied by a general feeling of emptiness: What are we going to do now about our lives?

In the second step, we fill that emptiness with thoughts that give us peace. This can be done in various ways, such as during retreats in a calming environment or just by studying relevant material.[96]

The goal is a restructuring of our lives where we become content and at peace with ourselves and the people around us, and where we eliminate dread and fear about the future. Eastern religions such as Daoism and Buddhism[97] contain impressive ideas for the second step.[98]

———————————

We move on to Part III, in which we use models to investigate philosophical questions and statements.

Part III

Confusion

13

Nonsense

This part of the book is different from Parts I and II. There we examined how subconscious and conscious models in the brain impact our lives. Here we rely on these models to clarify the use of language.

Such clarification is badly needed in our world. The Internet and social media often use language in a confusing manner, either unintentionally or on purpose.

This chapter and the next one describe a method that achieves the desired clarity. The subsequent chapter applies it to the long-standing philosophical question "Do we have free will?"

Let's begin. Chapter 12 may have given the erroneous impression that religions are the only way in which models of the world have been expanded.

Actually, our models of the world already contain a number of abstract concepts that have no counterpart in nature. Examples are the notion of beauty, essence, freedom, independence, knowledge, quality, reliability, and time.

We use each of these abstract concepts in diverse settings. In the spirit of this book, we say that each concept is employed in different conscious models.

For example, the word "quality" is used in varied statements such as "Parents must spend quality time with their children," "This is a high-quality clock," and "Today's air quality is at level red." These statements are part of conscious models covering parenting, production control, and air pollution.

———————

Philosophers often investigate abstract concepts independently of their specific use. An example question is "What is the essence of quality?"[99]

The philosopher Ludwig Wittgenstein (1889–1951) was troubled by such questions. In an effort spanning decades,[100] Wittgenstein determined that most of these questions do not make sense and thus are *nonsensical.*[101] In the quote below, the term "language-game" refers to any typical situation where a given word is used in everyday life.[102]

"When philosophers use a word—'knowledge', 'being', 'object', 'I', 'proposition', 'name'—and try to grasp the *essence* of the thing, one must always ask oneself: is the word ever actually used in this way in the language-game which is its original home?—

"What we do is to bring words back from their metaphysical to their everyday use."

Language-games are the core idea of Wittgenstein's scheme[103] for resolving philosophical questions. He clarified numerous philosophical questions using that approach. In our opinion, these a-chievements make him the greatest philosopher of the 20th century.

———————

The next chapter interprets Wittgenstein's scheme using the notion of subconscious and conscious models. We call the result *Wittgen-*

stein's method. With its aid, we show that the question "What is the essence of quality?" is nonsensical and cannot be answered.

We employ Wittgenstein's method almost daily to cut through the clutter and fog produced by nonsensical use of abstract concepts. Such confusion is purposely created in the alternate world of Chapter 11.

14

Wittgenstein's Method

We demonstrate Wittgenstein's method[104] using the question "What is the essence of quality?" as an example.

First, we assemble a number of statements containing the word "quality."[105]

Second, we select a typical situation where each statement might be made.

Third, in our mind we expand each situation into a complete model. We imagine people, actions, and surroundings.

Fourth, we immerse ourselves into each model—we live in it, so to speak—and interpret the meaning of the word "quality."

For example, take the sentence "Parents must spend quality time with their children." It might occur during a counseling session for parents. The word "quality" here means that parents should make time to lovingly interact with their children without distraction.

We summarize that case below, followed by a number of additional sentences involving the word "quality."

- "Parents must spend quality time with their children."
  Situation: counseling session for parents
  Meaning: loving interaction without distraction

- "This is a high-quality clock."

Situation: evaluation of technical equipment
Meaning: craftsmanship

- "Today's air quality is at level red."
  Situation: smog control program
  Meaning: level of pollution
- "This painting is of high quality."
  Situation: tour of a museum
  Meaning: artistic achievement
- "The radio signal is of poor quality."
  Situation: selection of antenna
  Meaning: level of radio signal
- "We must enhance the quality of the negotiations."
  Situation: evaluation of negotiation strategy
  Meaning: cleverness of negotiating tactic
- "Please assess the auditor's quality of work."
  Situation: review of hiring of auditor
  Meaning: accuracy of numbers

You may think that we have enough cases. Not so. Wittgenstein advises that we collect an abundance of instances to become aware of the full range of possibilities. We see the reason for this recommendation in a moment. So please bear with us as we go on.

- "We must check the quality of buildings."
  Situation: after-disaster analysis
  Meaning: safety of buildings
- "This affects the quality of the printed copies."
  Situation: newspaper production
  Meaning: percentage of good copies
- "Let's improve the quality of justice and police cooperation."
  Situation: review board for social justice
  Meaning: fairness for minorities
- "Consider the high quality of these universities."
  Situation: application for admission
  Meaning: level of instruction

- "You should improve the quality of this report."
  Situation: police training class
  Meaning: accuracy and style of report
- "The quality of child care is most important."
  Situation: selection of child care
  Meaning: attention paid to the child

In the fifth step of processing, we collect the various meanings in a list.

- loving interaction without distraction
- craftsmanship
- level of pollution
- artistic achievement
- level of radio signal
- cleverness of negotiating tactic
- accuracy of numbers
- safety of buildings
- percentage of good copies
- fairness for minorities
- level of instruction
- accuracy and style of report
- attention paid to the child

In the sixth and final step, we go over the list and try to identify commonalities.

There are some cases of similar meaning.

For example, "level of pollution," "level of radio signal," and "percentage of good copies" measure something that goes up with increasing quality.

As a second example, "loving interaction without distraction" and "cleverness of negotiation" could be claimed to have in common an intense human involvement.

But when we consider all cases simultaneously, we arrive at the conclusion that there is no essence. Hence the question "What is the essence of quality?" is nonsensical.

In general, one may describe Wittgenstein's method as follows.

- Assemble a large number of statements that are relevant for the question or claim to be investigated.

- Establish situations where the statements may occur.

- Enlarge the situations to complete models that one can dive into and live in.

- Enter into these models and obtain meanings that are relevant for the question or claim.

- Ponder these meanings to arrive at a convincing conclusion.

We emphasize that the above description of Wittgenstein's method is not to be viewed as a rigid set of rules, and that we modify or eliminate steps as needed. But we always retain the core idea. We construct models and live in them to obtain meanings.

We should mention that Wittgenstein's original scheme doesn't utilize *models of the world* and instead invokes detailed thoughts about *settings of the world*. The difference may seem trivial, but becomes important when we simultaneously consider subconscious and conscious models.

Wittgenstein also omits final conclusions such as "The question 'What is the essence of quality?' is nonsensical." Instead, he leaves it to the reader to absorb ideas and thoughts, and poses questions to prompt thinking about particular aspects. By that process the reader is to obtain the sought-after understanding.[106]

In contrast, we have explicitly written down the meaning of the word "quality" for each example. When the reader absorbs and compares these diverse meanings, it becomes clear that there is no common notion one could declare to be the essence of quality.

The next chapter examines the much more complicated philosophical question "Do we have free will?"

15

Free Will

For hundreds of years[107] the brain was considered a machine that converted input signals into output decisions. Since a machine turns identical input into identical output, that view implied that we act like robots controlled by a computer program. Is that the case, or do we have freedom of choice?

Free will is defined as the ability to select from different possible courses of action unimpeded.[108] The question about our power to choose is then "Do we have free will?"

The inquiries into that question started at least 2,500 years ago and continue today unabated.[109]

How is it possible that this question has not been resolved? Could it be that the question is nonsensical due to a misconception of the brain's role?

We will show that this is indeed the case. The main tool is Wittgenstein's method.

The conclusion is different from claims arguing via brain science that we do not have free will.[110]

———————————

We begin with some results of brain science. The brain is a dynamic organism, reconfiguring itself, absorbing information, and adjusting to changes of the body. Here are some important aspects:[111]

- The brain allocates capacity for the various processes for hearing, sight, smell, taste, and touch, and for control of the various components and operations of the body.

  The processes compete with each other for space since there is only a finite amount of real estate, so to speak. Hence there is no visual brain, or auditory brain, or sense-of-touch brain. It is a generic machine that configures itself for various tasks.

- After birth, the brain gradually learns to make sense of incoming signals and produce appropriate output. For example, it learns how to grab items, interpret images, understand speech, walk, and talk.

  The learning process continues throughout life. Here are some complex examples where signals for one sense are externally converted to signals of another sense:

  For a deaf person, sound is sometimes converted externally to touch—that is, auditory signals become pressure signals on the skin. After a while, the person "hears" via the skin.

  For a blind person, images are sometimes converted to pressure on the skin. After some training, the person "sees" via the skin. When images are converted to electrotactile shocks across the tongue, the person eventually "sees" via the tongue. When images are converted to sound, a person begins to "see" via earbuds.

  This may sound crazy, but is not. For the case of "seeing" with the tongue, brain imaging has shown that the signals received by the tongue are processed by an area of the brain that normally handles visual motion.

- The brain changes itself in complicated ways—just look at Chapters 3-8.

- The brain copes with misfortune—such as a person going blind, losing hearing, or losing a limb—by reallocating no-longer-needed capacity to existing functions. In exceptional cases, it can even reconfigure itself when half of the brain isn't present at birth or had to be removed at a very young age to combat an incurable disease.

Chapters 3-8 describe how conscious thought can influence subconscious models. The related theory is *cognitive behavioral therapy* (*CBT*). That idea captures part of an extensive process connecting subconscious and conscious models.

In one direction, conscious actions and decisions affect subconscious models, in agreement with CBT. We are not aware of these changes, but experience the effects later—for example, when the severity and frequency of panic attacks decreases.

In the reverse direction, subconscious models produce feelings, trigger thoughts, and initiate actions that affect conscious models. We are aware of these changes.

The change process isn't just happening sporadically. Rather, the exchanges should be viewed as two rivers of information flowing in opposite directions.

We parse the flow of those rivers into distinct segments when we look at specific situations. Let's do this for an example case where Sophia of Chapter 3 copes with a panic attack.

- Sophia climbs a staircase.

- A subconscious model determines that the physical effort is excessive and threatens the well-being of the body. It outputs a feeling of dizziness to the conscious level.

  Since Sophia has had therapy about panic attacks, another subconscious model outputs, "Most likely the dizziness is just a panic attack. Use the tools you have acquired to combat this."

- A conscious model of Sophia processes the dizziness and the message about the tools.

  She thinks, "I feel the dizziness, but surely this isn't a problem with my body, but another panic attack. Let's pause for a moment and then continue taking one step at a time while I firmly hold on to the rail."

She pauses, then climbs at a measured pace. The dizziness persists. But when she reaches the top of the stairs, the dizziness disappears.

She thinks, "Great! This definitely was a panic attack. I handled it well. My way of coping is working."

- Sophia's thoughts of success flow to the subconscious level and change the models triggering dizziness and the advice. The dizziness model is less likely to be activated, while the advice model is more likely to issue the message.

The example suggests the following general scheme.

- We are in some situation.
- Subconscious models process information about the situation and output feelings, trigger thoughts, and initiate actions that we become aware of. There are also actions we are not aware of; an example is an increase of blood pressure when anger rises.
- Conscious models process the feelings, thoughts, and actions, and come up with conclusions and decisions.

  The conclusions flow back to the subconscious models. They may also result in a change of the conscious models; for example, when they prove a conscious model to be wrong.

- Subconscious models adapt and change based on the conclusions received from the conscious level.

Here are a number of additional examples. Each case summarizes a variety of specific instances with particular twists and features.

You may want to imagine yourself to be in these instances and experience them, as demanded by Wittgenstein's method.

The "we" in these examples is not us, but just some person. We use "we" so that the examples are engaging and not just technical narration.

- *Situation*: In the morning, we meditate about anger as described in Chapter 7.

In the afternoon, we have a business meeting where we are unfairly criticized.

Subconscious model: The verbal attack triggers the thought "This is an opportunity to learn more about anger management."

Conscious model: The thought induces us to calmly react to the criticism. We also congratulate ourselves: "Well done! We got better at anger management."

Subconscious model: The final thought reinforces the subconscious model for anger management.

- *Situation*: We hike in the mountains. After two hours, we feel tired.

  Subconscious model: The feeling of fatigue and the thought "This may be an overly cautious evaluation" are pushed simultaneously to the level of consciousness; see Chapter 4.

  Conscious model: We consider that we have hiked for a short period. Rest surely isn't needed. Hence we decide to go on. At the same time, we rethink the arguments proving fatigue to be a feeling.

  Subconscious model: Renewed thought about fatigue reinforces the subconscious model claiming fatigue to be an overly cautious evaluation.

- *Situation*: We watch a movie with a dramatic human story of compassion.

  Subconscious model: The movie causes the brain to become sympathetic to misfortunes of others.[112]

  Seeing a homeless person on the way to the parking lot, a strong feeling wells up. We want to help.

  Conscious model: We decide to give him a generous amount of money. We think, "Great that I could help this person."

  Subconscious model: The concluding thought reinforces the model bringing up sympathetic feelings for people in dire need.

- *Situation*: We learn that the father of a colleague has died. We feel pressure to go the funeral home and express our condolences. But we are afraid to be confronted with death.

Subconscious model: The thought comes up that going to the funeral home and consoling the colleague may actually help us cope with the fear of death.

Conscious model: We decide to go to the funeral home. As we go home, we think, "Good that I did this and consoled the colleague. Encountering death is part of life."

Subconscious model: The final thought reinforces the model producing empathy and lessens the perceived threat posed by death.

Let's look at more complex and emotionally laden examples.

- *Situation*: We enter a supermarket to shop. A wonderful smell of baked goods greets us. There is a display with cakes.

  Subconscious model: Warm feelings and guilt surface: We had these cakes before, and they were wonderful. The guilt stems from the fact that we are overweight and should avoid cakes. There is one additional thought. We have been able to resist this temptation.

  Conscious model: We decide to forego the cakes. As we leave the supermarket, we think, "Good that I skipped the cake. It isn't really essential for my life."

  Subconscious model: The final thought strengthens the model saying that we are able to resist the temptation.

- *Situation*: We want to fetch the mail. It sounds easy. Just walk to the mailbox and back to the house. But it isn't easy for us.

  Subconscious model: A fear surfaces as we walk to the mailbox. Will our heart be able to handle this? But another thought comes up, too. We have a tool to cope with this fear.[113]

  Conscious model: We tell ourselves that the fear is real, but that it isn't based on realistic arguments. With these thoughts we continue to walk to the mailbox. Back in the house, we record another successful day of coping with that fear.

  Subconscious model: The success reduces the strength of the model claiming we are overstressing our body, and strengthens the model claiming we are okay walking outside.

- *Situation*: We committed an awful deed in the past. Only one other person knows about it. He has been blackmailing us and just called to demand more money.
  Subconscious model: The call triggers fears: We will go to jail, and our partner will find out; or we will be blackmailed forever.
  Conscious model: We decide to pay one more time. A dejected thought comes up: "How is this going to end?"
  Subconscious model: The model producing the feeling of helplessness is reinforced.

- *Situation*: We have stage IV cancer and will die in less than three months.
  Subconscious model: Every day, fears surface. We will die, our partner will not be able to cope financially, and the baby we adopted six months ago will not be properly cared for.
  Conscious model: We constantly suffer from these fears. Then we invoke a mantra: "One day at a time." It calms the turbulent thoughts. We finally think, "This is very difficult, but I am able to manage this awful situation."
  Subconscious model: The final thought strengthens the model that produces a feeling of calm.

We finish with some examples where subconscious models don't just output feelings and trigger thoughts, but also initiate actions.

- *Situation*: A piano virtuoso practices nine hours each day.

  Today she performs Rachmaninoff's Piano Concerto No. 3.
  Subconscious model: Subconscious models operate the fingers of her hand with lightning speed, receive the sound, and send feelings about tone and flow of the music to the conscious level.
  Conscious model: She tracks and evaluates the feedback from the subconscious models and adjusts speed and emphasis accordingly.
  Subconscious model: The adjustments of speed and emphasis are stored in the subconscious model for future performances.

- *Situation*: We are on vacation. During the past week, we have driven the same route every day from the hotel to the beach.
  Subconscious model: Today we are halfway to the beach. There is no other traffic. A subconscious model for the route causes us to steer a little bit to the left.
  Conscious model: We are surprised. "Why did we steer to the left?" Two seconds later a pothole comes into view. It would produce a nasty bump. But due to the correction two seconds ago, the car straddles the pothole without any adjustment required.
  "Hey, well done!" we congratulate ourselves to the performance of the subconscious model.
  Subconscious model: The praise reinforces the retention of the maneuver for subsequent trips to the beach.

- *Situation*: We have driven on the interstate highway for several hours. There is little traffic.
  Subconscious model: A subconscious model keeps the car in the right lane.
  Conscious model: We suddenly realize that we have no idea how we got to the current position. There is no memory for the last two minutes or so. It's as if we had engaged an autopilot and closed our eyes.
  Decision: "We've got to pay attention! This is really dangerous. Let's stop at the next rest area for a break."
  Subconscious model: A subconscious model is on alert looking for the next rest area sign. When it shows up, the model generates a feeling of relief. We will be able to rest and continue safely.

The cases demonstrate that the subconscious and conscious models of the brain change *continuously*. Even two identical decisions made some time apart—say a purchase of some item made yesterday and again today—are produced by different brains and hence need to be separately analyzed to identify underlying feelings and considerations. Indeed, each decision, by itself, changes the brain!

Let's return to the definition of free will as the ability to choose between different possible courses of action unimpeded.[114]

Can we characterize when the decision process at the conscious level is *impeded* or *unimpeded* by the feelings, thoughts, and actions produced by the subconscious models?

The answer is: No. Instead, we might say that an intricate *interaction* of the two types of models results in decisions. This includes that some decisions acknowledge or modify actions already initiated by subconscious models.

We represented the interaction earlier by two rivers flowing in opposite direction.

Here is an analogy. Let's consider the painting of pictures as a combination of color selection and brushwork. In the analogy, the color selection corresponds to the results of subconscious models, and the brushwork to the results of conscious models.

We cannot think of one without the other.

For example, eliminating color selection is like reproducing Monet's famous Impressionist paintings of the Rouen cathedral[115] as black-and-white images. The more than thirty glorious paintings—done at different times of day and year to capture various moods produced by sun and sky—would lose their soul.

One may argue that conscious models may be impeded by yet-to-be-defined external factors. What could these factors be?

For an answer, we go back to the example cases and check where and how external factors might be present.

When Sophia deals with a panic attack producing dizziness, how could her choices for coping with the attack be impeded? Maybe by some fear that she may be wrong in her analysis and is dizzy because her heart is failing? But this would be just another aspect of the situation, already captured by the notion that subconscious models supply feelings and trigger thoughts.

Indeed, as we dive into Sophia's case, we realize that there are *no* external factors. All factors influencing the decision can be represented by feelings and thoughts triggered by subconscious models.

These arguments rule out external factors for just one example. But going back and working through all the other examples, it becomes obvious that everything influencing the conscious models comes from subconscious models. There simply aren't any external factors.

The absence of external factors can also be argued as follows.

Everything enters the conscious models of the brain via some subconscious models. Indeed, information received by hearing, sight, smell, taste, and touch first enters the brain at the subconscious level. The same applies to information about our body, such as the level of oxygen, CO_2, and sugar in our blood.

In all cases, subconscious models initially process the information. In some cases, the entire output of the models is at the level of subconsciousness. An example is an increase of the pulse when the oxygen level in the blood drops. But generally the output includes feelings, thoughts, and actions that appear at the level of consciousness.

Hence, external factors directly influencing conscious models do not exist.

We conclude that one cannot talk about impeded or unimpeded decisions. Subconscious and conscious models that are an integral part of the decision process establish what can, should, or must be done or avoided. These models are in flux; indeed, the entire brain changes all the time.

The question of free will made sense when the brain was considered a fixed machine accepting inputs and producing outputs. But the brain established by modern science isn't such a machine, and the terms "impeded" and "unimpeded" do not apply to the brain's choices.

We conclude that the question "Do we have free will?" is non-sensical. We should emphasize that the above discussion does not constitute a *proof* of that conclusion, in the sense that theorems are *proved* in mathematics. But there is *very strong evidence* for its correctness.

————————

How is this result related to presence or absence of choice implicit in the following sentences?

- Attendance of the lecture is mandatory.
- Answers to the optional questions earn extra credit.
- She remained single by choice.
- We need a volunteer for this project.
- Why did he do this?
- That was a poor decision.
- No man is an island.
- Don't!

Over centuries, sentences of that kind motivated investigation whether we have decision-making freedom. The definition of free will as the ability to choose between different possible courses of action unimpeded[116] seemed an appropriate concept for that quest.

It turns out that brain science has upended the traditional concept of reasoning and decision-making. The new results do not invali-date the definition of free will as a certain ability, since definitions cannot be argued about. But the results show that the definition is not useful, and that the question "Do we have free will?" based on that definition is nonsensical.

————————

The laws of a country have many purposes, such as keeping the citizens safe from criminals, protecting property rights, and assuring fairness of elections. When a court of law has determined that a person has violated one of these laws, a judgment of liability or punishment makes sense only if the person had a choice.

A long time ago, such judgment generally assumed that the person had free will. That is, the person was able to choose between different courses of action unimpeded.

That universal assumption was gradually whittled down. For example, it was found that some persons had a diminished mental capacity; had been abused by parents or spouses, and thus brought to a breaking point; or had a chemical imbalance in the brain.

The changes were modifications of the concept of free will.

The conclusion drawn here that the question "Do we have free will?" is nonsensical, is more radical. It implies that the notion of free will generally isn't useful.

If we adopt that conclusion, we must create a concept different from the current notion of free will. It must respect the results of brain science and retain the notion that we are somehow responsible for our choices.

What might that concept be?

We don't have a reasonable answer, but suggest that Wittgenstein's method be used to weed out inappropriate proposals.

———————————

We have reached the end of our exploration of subconscious and conscious models. The epilogue has some final thoughts.

Part IV

Epilogue

We have taken a tour of some subconscious and conscious models that our brain builds to understand and cope with the world and the body.

The details of the models are consistent with the results of brain science. If the research in brain science proceeds in the expected way, we should be able to update the postulated models as new results become available. The condition "in the expected way" is crucial. For example, a new notion of consciousness might call for modifications beyond a simple refinement.

On a personal note: The trip across the landscape of brain science has been like a dream visit to a new and wonderful land. The journey has clarified a number of issues for us and has filled us with humility: What a wondrous world we live in!

Notes

Throughout, "Wikipedia" refers to the English version. All sources and links were accessed for final verification in March 2021.

Chapter 1 Introduction

1. Throughout "brain" refers to the organ consisting of cerebrum, cerebellum, midbrain, pons, and medulla. See Wikipedia "Human brain."

2. [Eagleman, 2020].

3. [Grafton, 2020].

4. [Conant and Ashby, 1970] establishes the fundamental mathematical result that "every good regulator of a system must be a model of that system." The result has the corollary that "the living brain, so far as it is to be successful and efficient as a regulator for survival, *must* proceed, in learning, by the formation of a model (or models) of its environment." [emphasis in the original]
Thus, the brain *must* form models to be effective.

5. See Wikipedia "Human brain."

6. [Hawking and Mlodinow, 2010].

7. [Kahneman, 2011].

8. The term "essentially inaccessible" means that we cannot access these models by conscious thought. Brain science is now able to pinpoint the location where some subconscious models reside. Surely we will gain additional insight in the future.

9. The theory of *physical intelligence* deals with such cases compre-

hensively; see [Grafton, 2020].

10. Precise prediction of the positions of stars and planets was first accomplished in 1627 by Johannes Kepler (1571–1630); see [Kepler, 1627].

11. See Wikipedia "Orbit," "List of future astronomical events,"and "Timeline of the far future."

12. [Hawking and Mlodinow, 2010].

13. [Rosling et al., 2018] puts forth the notion that the world is in better shape than typically depicted, and that the people of the earth *will* work out solutions.
For example, the Bill & Melinda Gates Foundation is proving that health care and education for young women not only results in economic success, but also reduces the birthrate by spaced pregnancies; for details, search the Internet for "Women's Economic Empowerment Gates Foundation."

Chapter 2 Things We Will Never Learn

14. [Cicero, 43].

15. Wikipedia "Checker shadow illusion." By Edward H. Adelson, own work, and vectorized by Pbroks13. https://en.wikipedia.o rg/wiki/Checker_shadow_illusion#/media/File:Checker_shadow_i llusion.svg. Licensed under CC BY-SA 4.0 via Commons.

16. Wikipedia "Checker shadow illusion." By Edward H. Adelson, own work, and vectorized by Pbroks13. https://en.wikipedia.o rg/wiki/Checker_shadow_illusion#/media/File:Grey_square_opti cal_illusion_proof2.svg. Licensed under CC BY-SA 4.0 via Commons.

17. Source: "Shepards tables illusion." By tables - Own work, CC BY-SA 4.0, https://commons.wikimedia.org/w/index.php?curid= 64967958.

18. Source: "Surfaces of Shepard's tables." Extracted by K. Truemper from work by tables, CC BY-SA 4.0, https://commons.wikime dia.org/w/index.php?curid=64967958.

19. See Wikipedia "Shepard tables" for details about this amazing construction by Roger N. Shepard.

Chapter 3 Psychotherapy

20. [Burns, 2020] provides not only a clear introduction, but is sufficiently detailed for readers who are looking for solutions they can implement by themselves. For an overview of CBT, see Wikipedia "Cognitive behavioral therapy."

21. Dr. M. Steigleder, who specializes in psychotherapy of anxiety and depression disorders, helped with the assembly of the fictitious story.

22. The classic [Eliot and Breo, 1989] helps with those changes.

23. [Beck et al., 1979], [Burns, 2008], and [Burns, 2020].

24. See Wikipedia "Bibliotherapy." For the case of depression, see [Burns, 2008] and [Burns, 2020].

Chapter 4 Fatigue

25. Craig Glenday, Editor-in-Chief at Guinness World Records, reports on https://www.quora.com/What-is-the-longest-distance-a-person-has-walked-in-one-go:
"Georges Holtyzer of Belgium walked 673.48 km (418.49 miles) in 6 days 10 hr 58 min, completing 452 laps of a 1.49 km (0.92 mile) circuit at Ninove, Belgium, from July 19 to July 25, 1986. He was not permitted any stops for rest and was moving 98.78 percent of the time."

26. p. 209 [Grafton, 2020].

27. p. 210, 211 [Grafton, 2020].

28. p. 213 [Grafton, 2020].

29. See Wikipedia "Paavo Nurmi."

30. p. 214 [Grafton, 2020].

31. [Gibson et al., 2013] describes three stages of collapse.

- During the early stage—the "Early Foster" collapse position—the runner exhibits unstable gait and lowers the head.
- The gait deteriorates to a shuffle in the "Half Foster" collapse position, with head parallel to the ground.
- In the final stage—the "Full Foster" collapse position—the run-

ner crawls on the ground on elbows and knees and finally collapses before or after reaching the finish line.

The reference conjectures that the collapse positions are indicative of a final, likely primordial, protective mechanism.

Chapter 5 Breathless

32. The first part of this chapter is based on Wikipedia "Chronic obstructive pulmonary disease."

33. [Nestor, 2020].

34. See Wikipedia "Medulla oblongata."

35. Source: "Medulla of brain" by OpenStax - https://cnx.org/co ntents/FPtK1zmh@8.25:fEI3C80t@10/Preface, CC BY 4.0, https: //commons.wikimedia.org/w/index.php?curid=30147954.

36. The discussion is based on [Nestor, 2020].

37. p. 76 [Nestor, 2020].

38. The extreme case is hyperventilation, where more CO_2 is eliminated by the lungs than produced by the body. See Wikipedia "Hyperventilation."

39. p. 81 [Nestor, 2020].

Chapter 6 Motion

40. See Wikipedia "Polar curve (aerodynamics)."

41. See Wikipedia "List of domesticated animals."

42. See Wikipedia "Human evolution."

43. See Wikipedia "Walking."

44. See Wikipedia "Walking."

45. p. 190 [Grafton, 2020].

46. [Selinger et al., 2015]. See also p. 192 [Grafton, 2020] and the section on Energetics of Wikipedia "Preferred walking speed."
The total energy consumption per mile is called *gross cost of transport*. The Wikipedia entry also discusses an alternate model that

minimizes the incremental energy used beyond the basal metabolic rate. The resulting optimal speed is a bit lower than the typical speed of healthy pedestrians. Hence, the model based on gross cost of transport cost seems more appropriate.

There is another argument for that choice. Suppose thousands of years ago a hunter/gatherer hiked to a new area for hunting animals and collecting plants. The optimal speed minimizing gross cost of transport would be the best way to assure survival of the trip. Thus, evolution likely installed that choice in the subconscious brain.

Chapter 7 Rest

47. In 2018, more than 166 million adults played video games in the U.S. https://review42.com/video-game-statistics/.

48. The idea of the news fast sounds crazy, doesn't it. But it is wonderful. When we get back into our routine, we realize how much of that flood of daily information is useless drivel that clutters up our minds.

49. https://en.wikipedia.org/wiki/Meditation.

50. For a number of books on mindfulness, including the classic [Kabat-Zinn, 1990], see Wikipedia "Jon Kabat-Zinn."

51. Various mindfulness training programs can be downloaded free of charge: http://www.freemindfulness.org/download. In the discussion of this chapter, we specifically mean the program "Forty five minute body scan." It requires 45 minutes, as stated in the title. But later, when we have become good at the body scan, we can reduce it to something like 20 minutes, done each day.

52. The 14th Dalai Lama has done much to bring these thoughts to the Western World. Search the Internet for "books by the Dalai Lama." The arguments cited here are taken from [His Holiness The Dalai Lama and Cutler, 1998].

Chapter 8 Training

53. See Wikipedia "Piano Concerto No. 3 (Rachmaninoff)."

54. See Wikipedia "Otto Lilienthal."

55. See Wikipedia "Wright brothers."

Chapter 9 Medicalization

56. See Merriam-Webster dictionary "medicalization."

57. See Wikipedia "Drapetomania."

58. "Lesions on the back of an enslaved African from Mississippi" by Mathew Brady - This media is available in the holdings of the National Archives and Records Administration, cataloged under the National Archives Identifier (NAID) 533232., Public Domain, https://commons.wikimedia.org/w/index.php?curid=284064.

59. See Wikipedia "Dysaesthesia aethiopica."

60. See Wikipedia "Female hysteria" for a summarizing discussion and [Tasca et al., 2012] for details.

61. [Caughey et al., 2014].

62. See Wikipedia "Caesarean section."

63. See Wikipedia "Hormone replacement therapy" and "Women's Health Initiative."

64. See Wikipedia "Hormone replacement therapy."

65. We once heard the receptionist in a doctor's office exclaim, "It's party time!" when a Pharma marketing representative arrived.

66. Search the Internet for "changing lifestyle."

Chapter 10 Misevaluation

67. p. 9 [Meadows et al., 1972].

68. [Meadows et al., 1972].

69. See also Wikipedia "Club of Rome."

70. [Rosling et al., 2018].

71. The factors 44 = US $3,980/China $90 and 110 = US $11,000/China $100 are derived from p. 43, 44 [Meadows et al., 1972].

72. The factor 38 = US $36,433/China $959 is derived from Wikipedia "List of countries by past and projected GDP (nominal) per capita," section "IMF estimates between 2000–2009."

73. The factor 6 = US $65,760/China $10,410 for 2019 is derived from Wikipedia "List of countries GNI (nominal) per capita." GDP and GNI are sufficiently close to permit the comparison with the data for the year 2000.

74. p. 53 [Rosling et al., 2018]. Extreme poverty is defined by income of less than $2/day.

75. See Wikipedia "Union Electric Company."

76. The story is based on private communication with a mid-level manager of Union Electric in 1965.

77. The subsequent part of the chapter builds upon notes supplied by Dr. A. Rajan and related discussions.

78. See Wikipedia "Executive compensation" and "Executive compensation in the United States" for an overview. Details are provided by [Murphy, 2012]. For the most recent rules covering executive compensation, search the Internet for "Research Handbook on Executive Pay."

79. [Murphy, 2012].

80. In technical terms, the borrowing of funds leverages up the company, and stock price becomes more like an option. Effectively, top management increases the value of the option as well as the likelihood of bankruptcy. In the language of gambling, the expected gain for top management is huge.

81. See Wikipedia "Pension crisis."

82. See https://www.epi.org/publication/ceo-compensation-surged-14-in-2019-to-21-3-million-ceos-now-earn-320-times-as-much-as-a-typical-worker/.

83. See the chart of the Board of Governors of the Federal Reserve System at https://www.federalreserve.gov/releases/z1/dataviz/dfa/distribute/chart/.
Contributing to the rise in wealth were significant tax reductions that favored the wealthy and were of little or no benefit to the lower middle class and the poor.

84. See the chart of the Board of Governors of the Federal Reserve System at https://www.federalreserve.gov/releases/z1/dataviz/dfa/distribute/chart/.

Chapter 11 An Alternate World

85. See Wikipedia "Theory of relativity."

86. See Wikipedia "List of federal judges appointed by Donald Trump."

87. See Wikipedia "2021 storming of the United States Capitol."

88. See Wikipedia "List of fact-checking websites."

89. The Wikimedia website https://stats.wikimedia.org has detailed statistics.

Chapter 12 Expanded Worlds

90. See Wikipedia "Prehistoric religion" and "Paleolithic religion."

91. Source: Big Bang model by NASA/WMAP Science Team - Original version: NASA; modified by Cherkash. https://commons.wiki media.org/w/index.php?curid=11885244. Public Domain.

92. See Wikipedia "Galaxy." Two trillion = $2 \cdot 10^{12}$.

93. There is a lurking threat, though, to supposedly untestable claims of current religions. Some time in the future, brain science may determine that the claims are rooted in subconscious constructions and hence are man-made.

94. [Dor-Ziderman et al., 2019].

95. See Wikipedia "List of religious populations."

96. [His Holiness The Dalai Lama and Cutler, 1998] is an example of books that extract wisdom from religious tenets.

97. See Wikipedia "Taoism" and "Buddhism." The word "Daoism" is the modern spelling of "Taoism."

98. The reader may wonder why we do not include Christianity and Islam—the world's most popular religions—in the list of potential sources. It seems that these religions so intricately motivate desirable behavior by the concepts of sin and damnation that it is difficult to extract other reasons.
The cited Eastern religions allow one to do so more easily, as demonstrated by [His Holiness The Dalai Lama and Cutler, 1998].

Chapter 13 Nonsense

99. The opening paragraph of Wikipedia "Quality (philosophy)" contains the statement "In contemporary philosophy, the idea of qualities, and especially how to distinguish certain kinds of qualities from one another, remains controversial."

100. See the "Ludwig Wittgenstein" entries of Wikipedia and Stanford Encyclopedia of Philosophy.
Books detailing Wittgenstein's life and accomplishments include [Monk, 1990], [Anscombe, 1971], [Hartnack, 1965], and [Fann, 2015]. The latter two books are particularly easy to follow. [Truemper, 2017] has a summary.

101. "Is Rome east of voltage?" is a simple example of a nonsensical question. The capital of Italy does not lie east of a concept of electricity. "Sun plus moon equals time" is another example. Addition of sun and moon cannot have time as a result. In both cases, the syntax is correct, but collectively the terms do not make sense.

102. Paragraph 116 [Wittgenstein, 1958].

103. The central publication for the scheme is [Wittgenstein, 1958].

Chapter 14 Wittgenstein's Method

104. The method is an interpretation of Wittgenstein's scheme of *language-games*, using the notion of models. The main publication is [Wittgenstein, 1958].

105. Translator programs providing example sentences may be helpful for this task. An example is Linguee; see linguee.com.

106. Wittgenstein's early work [Wittgenstein, 1963] says that nonsense in philosophical statements can only be shown but not proved. But the book engages in exactly such discussion. Wittgenstein resolves this conflict by a famous ladder argument where the reader uses the book as a ladder to achieve a level of understanding, and afterwards discards it.
In later work, Wittgenstein also avoids explicit claims that certain statement are nonsensical. Instead, he supplies numerous arguments that convincingly demonstrate the flaws. Thus Wittgenstein's books are treasure troves of profound insight into numerous settings.

A good example is Wittgenstein's demonstration that Goethe's theory of color is untenable; see [Goethe, 1810] and [Wittgenstein, 1978].

Chapter 15 Free Will

107. The eminent physician, surgeon, and philosopher Galen (129–c. 210) identified 1,800 years ago the brain as the source of emotions and thoughts. See Wikipedia "Galen."

108. See Wikipedia "Free will."

109. See Wikipedia "Free will" and "Neuroscience of free will," as well as Stanford Encyclopedia of Philosophy "Free will." The Stanford summary has more than 18,000 words. An Internet search of "free will" in early 2021 produced more than 12 billion results.

110. Wikipedia "Neuroscience of free will" cites a number of references claiming that free will is an illusion. The discussion focuses on the subconscious and conscious decision-making that happens at a given point in time. The debate ignores the fact that conscious thinking influences subconscious models, as postulated by CBT and demonstrated in Chapters 3–8 and in this chapter. In particular, we can consciously influence today what subconscious models will do tomorrow. Psychotherapy interventions rely on this fundamental fact.

111. See [Eagleman, 2020] for details about the brain's performance.

112. [Zak, 2015]. The chemical Oxytocin triggers the sympathetic reaction.

113. See Chapter 3.

114. See Wikipedia "Free will."

115. See Wikipedia "Rouen Cathedral (Monet series)."

116. See Wikipedia "Free will."

Bibliography

[Anscombe, 1971] Anscombe, G. E. M. (1971). *An Introduction to Wittgenstein's Tractatus*. St. Augustine's Press.

[Beck et al., 1979] Beck, A. T., Rush, A. J., Shaw, B. F., and Emery, G. (1979). *Cognitive Therapy of Depression*. Guilford Press.

[Burns, 2008] Burns, D. D. (2008). *Feeling Good: The New Mood Therapy*. Harper.

[Burns, 2020] Burns, D. D. (2020). *Feeling Great: The Revolutionary New Treatment for Depression and Anxiety*. PESI Publishing & Media.

[Caughey et al., 2014] Caughey, A. B., Cahill, A. G., Guise, J.-M., and Rouse, D. J. (2014). Safe Prevention of the Primary Cesarean Delivery. *American Journal of Obstetrics & Gynecology*, pp. 179–192.

[Cicero, 43] Cicero, M. T. (43). *Phillipicae*. Part XII. Speech delivered at the beginning of March 43 BCE. Translation by C. D. Yonge. Lexundria, https://lexundria.com/cic_phil/12/.

[Conant and Ashby, 1970] Conant, R. C. and Ashby, W. R. (1970). Every good regulator of a system must be a model of that system. *International Journal of Systems Science*, vol. 1, pp. 89–97.

[Dor-Ziderman et al., 2019] Dor-Ziderman, Y., Lutz, A., and Goldstein, A. (2019). Prediction-based neural mechanisms for shielding the self from existential threat. *NeuroImage*, vol. 202.

[Eagleman, 2020] Eagleman, D. (2020). *Livewired: The Inside Story of the Ever-Changing Brain*. Pantheon Books.

[Eliot and Breo, 1989] Eliot, R. S. and Breo, D. L. (1989). *Is It Worth Dying For? A Self-Assessment Program to Make Stress Work for You, Not Against You.* Bantam Books.

[Fann, 2015] Fann, K. T. (2015). *Wittgenstein's Conception of Philosophy.* Partridge Publishing.

[Gibson et al., 2013] Gibson, A. S. C., De Koning, J. J., Thompson, K. G., Roberts, W. O., Micklewright, D., Raglin, J., and Foster, C. (2013). Crawling to the finish line: why do endurance runners collapse? Implications for understanding of mechanisms underlying pacing and fatigue. *Sports Medicine,* vol. 43, pp. 413–424.

[Goethe, 1810] Goethe, J. W. (1810). *Entwurf einer Farbenlehre.* German and English versions available at https://theoryofcolor. org/Theory+of+Color.

[Grafton, 2020] Grafton, S. (2020). *Physical Intelligence: The Science of How the Body and the Mind Guide Each Other Through Life.* Penguin Random House.

[Hartnack, 1965] Hartnack, J. (1965). *Wittgenstein and Modern Philosophy.* Methuen.

[Hawking and Mlodinow, 2010] Hawking, S. and Mlodinow, L. (2010). *The Grand Design.* Bantam Books.

[His Holiness The Dalai Lama and Cutler, 1998] His Holiness The Dalai Lama and Cutler, H. C. (1998). *The Art of Happiness.* Riverhead Books.

[Kabat-Zinn, 1990] Kabat-Zinn, J. (1990). *Full Catastrophe Living.* Dell Publishing.

[Kahneman, 2011] Kahneman, D. (2011). *Thinking, Fast and Slow.* Farrar, Straus, and Giroux.

[Kepler, 1627] Kepler, J. (1627). *Tabulae Rudolphinae (Rudolphine Tables).* https://archive.org/details/tabulaerudolphin00ke pl/page/n1/mode/2up. A book produced in 2014 contains the original Latin text and a German translation. It uses the fonts and graphics of the original book for both versions—an astonishing achievement. Title: *Die Rudolphinischen Tafeln.* Editor: Jürgen Reichert. Publisher: Königshausen & Neumann, 2014. See `https://www.amazon.de/Die-Rudolphinischen-Tafeln-J%C3%BCrgen-Reichert/dp/3826053524`.

[Meadows et al., 1972] Meadows, D. H., Meadows, D. L., Randers, J., and Behrens III, W. W. (1972). *Limits to Growth*. Universe Books.

[Monk, 1990] Monk, R. (1990). *Ludwig Wittgenstein: The Duty of Genius*. Penguin Books.

[Murphy, 2012] Murphy, K. J. (2012). Executive Compensation: Where We Are, and How We Got There. Available at `http://ssrn.com/abstract=2041679`.

[Nestor, 2020] Nestor, J. (2020). *Breath: The New Science of a Lost Art*. Riverhead Books.

[Rosling et al., 2018] Rosling, H., Rosling, O., and Rönnlund, A. R. (2018). *Factfulness*. Flatiron Books.

[Selinger et al., 2015] Selinger, J. C., O'Connor, S. M., Wong, J. D., and Donelan, J. M. (2015). Humans Can Continuously Optimize Energetic Cost during Walking. *Current Biology*, vol. 25, pp. 2452–2456.

[Tasca et al., 2012] Tasca, C., Rapetti, M., Carta, M. G., and Fadda, B. (2012). Women And Hysteria In The History Of Mental Health. *Clinical Practice & Epidemiology in Mental Health*, vol. 8.

[Truemper, 2017] Truemper, K. (2017). *The Construction of Mathematics – The Human Mind's Greatest Achievement*. Leibniz Company.

[Wittgenstein, 1958] Wittgenstein, L. (1958). *Philosophical Investigations*. Basil Blackwell; available at `https://drive.google.com/file/d/0Bw-duXxYihdvWVlFaUhzclY5Vmc/edit`.

[Wittgenstein, 1963] Wittgenstein, L. (1963). *Tractatus Logico-Philosophicus*. Routledge & Kegan Paul Ltd; go to `people.umass.edu/klement/tlp/tlp.pdf` for the German version and two translations into English.

[Wittgenstein, 1978] Wittgenstein, L. (1978).   *Remarks on Colour*. University of California Press.

[Zak, 2015] Zak, P. J. (2015). Why inspiring stories make us react: the neuroscience of narrative. *Cerebrum*, February, 2015.

Acknowledgements

We couldn't have written this book without extensive assistance in several areas. The key persons for technical advice, evaluation of chapters, and corrections, were M. Grötschel, W. R. Osterhoudt, A. Rajan, M. Steigleder, and T. Willmann.

I. Truemper and U. Truemper were patient editors.

The University of Texas at Dallas—our home institution—made essential resources available.

We thank all of them for their help.

K. T.

Index

www.ingramcontent.com/pod-product-compliance
Lightning Source LLC
Chambersburg PA
CBHW032008190326
41520CB00007B/401